Nature as Constructor

The purpose in writing this book was to give the greatest possible circle of readers an insight into phenomena in the vegetable and animal kingdom which lend themselves to an interpretation from the viewpoint of physics, technology and cybernetics, or are related directly or indirectly to technical development in human civilization and culture. What has brought the biologist and the physicist together as its authors was the fun of roving through border disciplines of biology and physics plus technology and of pondering on and sifting well-known material. A new orientation in scientific work—referred to as bionics—is a relatively recent development (about 1960). Its main objective has been to create while following the example of animate nature new kinds of highly effective machines and systems for the benefit of Man. For this bionics prepares the ground by systematically investigating the multiplicity of biological structures, forms and processes and the ways these are functionally interrelated. Examples of bionical research are given in the ensuing pages. However, the book has yet another aim in view. In our age of highly advanced technology when practically every child knows how a motor-car, a refrigerator or a television set works contemplating animate nature as a physical and technical phenomenon may we believe be the process most susceptible of awakening greater interest in, and perhaps more understanding for, our endangered natural environment.

In this sense our words are addressed less to the specialist than to the man in the street. While hardly all phenomena coming under the above head have been treated it is believed that a representative selection out of their abundance has been made. In a way we assume that many a reader has assimilated certain basic concepts of biology and physics as part of his general knowledge. If so, he will find it so much easier to gain insight into the rather complex interrelations and the integrated and technical struc-

tures and functions. Here the introductory chapters should be a help. We have tried to explain in these introductions as clearly as possible at least a few physical laws indispensable for grasping the individual examples. Specialized discussions and downright hypothetical arguments have been avoided wherever possible as being the domain of scientific literature.

Nature is not a constructor in the sense that the engineer is. In the last resort she is inimitable. Even so the technician should, more often than he has done hitherto, venture a glance at biological structures. He will hardly find ready-made recipes and solutions of his own technical problems but he may expect a variety of interesting hints. This is true of physical phenomena that can be explained by laws of physics and cybernetics much rather than static constructions. It is certainly no exaggeration to say that compared with the central nervous system of simple insects modern integrated micro-electronics is a poor humdrum affair. Of course, in nature everything happens correctly, in a "natural" way. In fact, however, mutation and selection, the two great artificers in changes of species of plants and animals, merely appear to be operating in a teleological way in the course of biological evolution. Consequently, organic structures are often a long way from corresponding to ideal concepts—even though in most organisms their fitness to survive has inevitably been improved by natural selection.

Animate nature as the great instructor for technology and—vice versa as a kind of backfeed—significant advances in biological research brought about by applying physical and technological viewpoints and data: it is this reciprocal action that the following pages seek to elucidate. Should our readers feel that the book has instruction as well as entertainment in the good sense of the word to offer the authors will be well content.

Berlin and Dresden, winter 1976   Klaus Wunderlich   Wolfgang Gloede

Klaus Wunderlich / Wolfgang Gloede

Nature as Constructor

ARCO PUBLISHING, INC.
New York

Translated from the German
by Dr Vladimír Vařecha

Drawings in the text
by Michael Lissmann, Leipzig and
Hans Haubold, Leipzig

Designed by Walter Schiller, Altenburg

Published 1981 by Arco Publishing, Inc.
219 Park Avenue South, New York, N.Y. 10003

Library of Congress Cataloging in Publication Data

Wunderlich, Klaus, Dr. rer. nat.
  Nature as constructor.

  Translation of Natur als Konstrukteur.
  Bibliography: p. 192
  Includes index.
  1. Animal mechanics.   I.   Gloede, Wolfgang, joint
author.   II.   Title.
QP303 W8613   1980   574 19'12   80-18311
ISBN 0-668-05102-7

Problems of Statics
in Plants and Animals

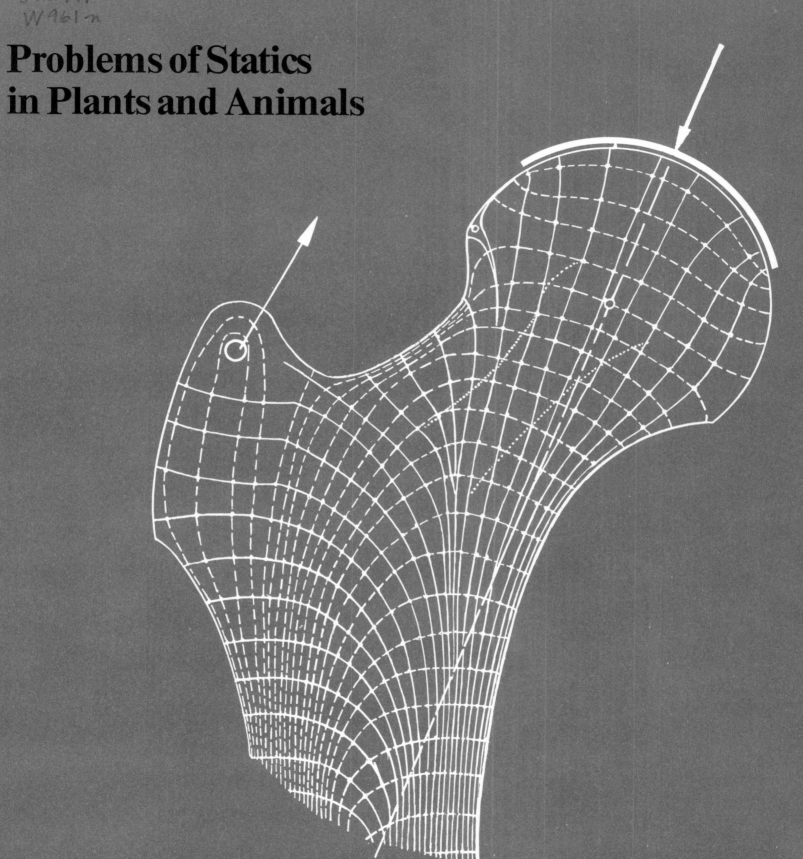

At the outset simple biological structures encountered by Man in his surroundings served him as models for building purposes. Those constructions with their manifold forms and functions had arisen in the course of phylogenesis through mutation and selection. It was in this way that highly effective and ultra-light structures gradually developed. Yet there are many outstandingly workable design principles in technology that had been developed by scientists and engineers for the most part without as much as a side-glance at models to be found in nature. These are all based on physical laws. It is becoming increasingly evident, however, that in the case of plants and animals in both macrosphere and microsphere design principles have been put into practice just in the sphere of statics which are equal, or even superior, to the artificially created ones. "Investigations of vegetable and animal kingdoms as to static principles have brought to light astonishing solutions which differ from the technical ones possibly not so much in the conception of the respective elements as in the extraordinarily harmonious cooperation of the elements in fulfilling a superordinate function." (cf. W. Nachtigall, 1971) Biostatic investigations therefore have not to result solely in the complete clarification of the connection between form and function of biological constructions; nevertheless, they can give the technicians useful suggestions for original solutions of technical problems.

Living structures, too, are subject to laws of physics. Only the demands made upon them here are quite different from those in engineering, while building materials used in the two fields are also widely different. Static technical structures, such as chimneys, television towers, skyscrapers, but also precision machinery and adjustment benches are expected to be as rigid as possible, and should neither bend nor vibrate. In natural design, on the contrary, almost everything is elastic and flexible, just think of trees and grasses, or of bones, cartilages and ligaments in the skeleton of a mammal. To be sure, in animate nature, too, if structural units are not to be destroyed their deformation must not exceed certain values. Deformations are engendered by forces which either derive from the weight of the structure itself (weight force equals mass times gravitational acceleration; in the ensuing paragraphs weight force will be referred to in a simplifying way as weight), or operate from outside. In structural units they give rise to tensile or compressive stress, bending, breaking, shearing or torsion. Principles of the lever and resolution of forces (parallelogram of forces) play an important part here.

Apart from the physical laws of statics, stability of a structure is substantially affected by the properties of the material. It is the elastic modulus of a material that applies here, which is established from the relative change in length $\frac{\Delta l}{l}$, arising during the process.

According to the Hooke law (Robert Hooke, 1635–1703, outstanding English physicist and contemporary of Newton, his discoveries include the cellular structure of the plant body) there is a certain load range for elastic materials, in which the relative modification in length is directly proportional to the tensile stress and the elastic modulus:

$$\frac{\Delta l}{l} = \frac{\sigma}{E}$$

[σ = tensile stress = tensile force divided by sectional area, E = elastic modulus].

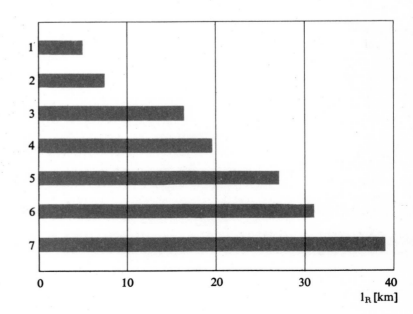

Paradoxically enough, highly elastic materials $\left(\frac{\Delta l}{l} \text{ large}\right)$ have a small elastic modulus $\left(\text{caoutchouc: } E \approx 0.1 \frac{\text{kgf}}{\text{mm}^2}, \text{ steel, on the other hand,: } E \approx 10^5 \frac{\text{kgf}}{\text{mm}^2}\right).$

Furthermore, the degree of the aptitude of the material is determined by tensile stress at the elastic limit and by that at the ultimate stress limit. Thickness of building materials is yet another quantity not to be underestimated by the designer. It determines the dead weight of his construction, which often greatly exceeds the weight of additional loading from outside (bridges, aerial masts, etc.). What is called breaking length clearly illustrates the influence of density upon static properties. This is the length at which a rod or wire hanging down vertically breaks under its own weight. Diagram 1 makes it clear how vastly superior are e.g. spider threads to structural steel in this respect. However, without regard to their slight density, the tensile strength of some vegetable and animal tissues is of a very high order, and with their much higher extensibility surpasses that of most metals.

Sclerenchyma strips (strips of strengthening tissue whose cells have thickened outer walls) of plants reach approximately the tensile strength of good sorts of structural steel $\left(\approx 20 \frac{\text{kgf}}{\text{mm}^2}\right).$

However, properties of materials are not the only factors of importance for the static and dynamic behaviour of supporting structures: the decisive aspect is the form and profile of structural elements. Fig. 2 shows a beam restrained in one direction and loaded with a weight at its free end. The elastic beam deflects while extending slightly above its centre line (tensile stresses) and becoming slightly compressed below it (compressive stresses). There is no deformation in the centre line itself (neutral fibre), which is stress-free. The beam could be hollowed out in the middle and its load-capacity would remain practically the same. Rod cross-sections where the same sagging occurs under the same loading are shown in Fig. 3. Material inventory of the last (d)

1   Breaking lengths l_R of some technical and natural materials. (The breaking length of a material is the length at which a vertically suspended rod or wire breaks under its own weight.)
1 Structural steel,
2 Vulcanized fibre,
3 Beechwood,
4 Aluminium alloy,
5 Spider's web,
6 Horn (feather),
7 Stretched glass fibre.

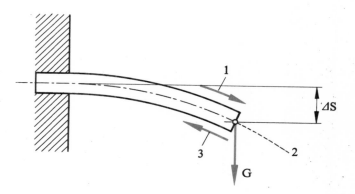

2   Bending of a unilaterally fixed beam by loading weight G.
1 indicates direction of tensile forces which stretch the beam in the upper range.

2 is the centre line (neutral fibre) of the beam in which no material deformation occurs. Compressive forces (3) bulge the beam in the lower range. Δs is deflection of the beam from normal position.

profile (tube) amounts to mere 25 per cent of that of the second (solid rod). Thus flexural rigidity essentially depends on the way the material is distributed round the neutral fibre; the material must be concentrated as far as possible on the edge of the profile. This accounts for the superiority of a tube over a solid rod. In deriving the formula from the rod's bending moment a proportionality is established to the material area elements in the profile cross-section and to the square of the distance of these surface elements from the neutral fibre. The product of the area and of the square of distance is called areal moment of inertia, its calculation being based on the integral calculus:

areal moment of inertia $= \int y^2 \, dA$

[y = distance, dA = differential area element].

This is analogical to the well-known mass moment of inertia which equals the product of the mass and the square of its distance from an axis of rotation (for instance, in a flywheel—which must possess a high moment of inertia to be able to counteract variations in speed—the largest part of its mass is concentrated on the circumference). High areal moment of inertia means high resistance against bending and buckling.

Hollow cylindrical bones of vertebrates and tubular grass-blades are examples of an exceedingly solid structure with the smallest possible material inventory. More will be said about both in separate chapters.

Finally, we wish to discuss another major aspect which is often left out of account when comparing natural structures among themselves, or with technical ones:

It is impossible simply to enlarge static constructions proportionally in all dimensions and expect a proportional increase in load-capacity. As already mentioned, dead weight itself constitutes an essential loading—frequently even the only one (trees, chimney stacks). When proportionally increased this weight grows with the third power of length, whereas the supporting sectional areas only with the second power. Thus with increasing height the ratio of sectional area to volume constantly decreases until eventually the structure breaks down under its own weight. If the load-carrying capacity required for adequate stability is to grow in proportion to the weight, then thickness d must grow more than proportionally to length l, —i.e. $d \sim l \cdot \sqrt{l}$ and not $d \sim l$. Bearing this law in mind will save us from erroneous interpretations in making comparisons between towers or smoke stacks and stalks of grass.

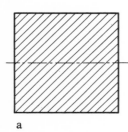 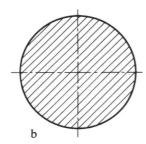 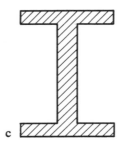 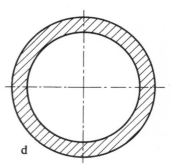

a   b   c   d

3   Various cross-sections of beams which when equally loaded by a weight G show the same deflection. Attention should be paid to the widely different inventory of material in the four sections. Material inventory: a 100%, b 103%, c 39%, and d 25%.

Flexural rigidity increases if the material is disposed more on the periphery as, for instance, in grass stalks or tubular bones.
After R. W. Pohl.

4 The giant floating leaves of *Victoria amazonica* have already stood a 75-kilogram load per leaf in experiment. The amazing carrying capacity is enhanced by a multitude of broad veins.

6 The engineer derives inspiration
from nature: the roof of the
Olympic Stadium in Munich.

7   Occasionally the extreme slenderness of long stalks has evoked the false idea of nature's superiority over comparable technical constructions.

Are Stalks of Grain Natural Skyscrapers?

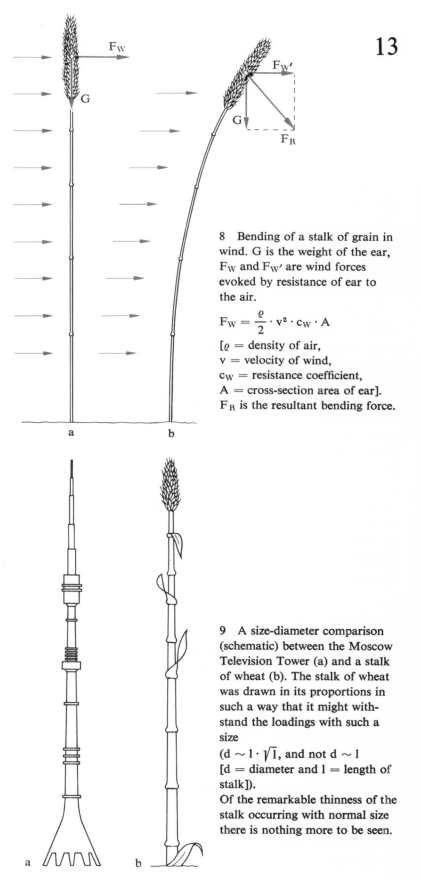

Who would not be delighted to recall the splendid sight offered by an early-summer cornfield in motion? If possible, one would be standing on an elevation, perhaps a moraine knoll rising 20 or 30 metres above the surrounding stretch of fields. The wind and the sun had transformed the yellow-green surface into an undulating sea whose millions of stalks were swinging in the rhythm of the moving air like waves running, smooth and irresistible, across the glittering expanse of the waters. Passing clouds threw blue fleeting shadow images upon it, and here and there rose mellow yellow pillars which soon dispersed again—tiny pollen from innumerable powdering ears rapidly wafted away by the wind. This is a view one can enjoy not only amidst one's native field of corn, it is no less beautiful in the blossoming fescue and feather grass steppe of the South Ukraine, in the East African savannahs, or the North American prairies.

In all those places our eyes rest on grasses, members of the family of flowering plants, the most important plants for mankind. With its 8,000 to 10,000 species this voluminous group populates all parts of our planet wherever vegetable life is possible. For many a man it is nothing short of a miracle to observe how so thin a hollow stalk, with the additional heavy load of ears or panicles, withstands the pressure of the strong wind without breaking unless prolonged heavy rains allied with stormy winds have forced the corn or ripe grass to "lie down". Consequently, a number of well-meaning writers have attempted to regard the natural static construction called "stalk of straw" as vastly superior to comparable technical structures. This tradition is one of a piece with the usual mistake concerning the phenomenal "slenderness ratio" of the stalk of rye which has been compared with a chimney stack—this has recently been pointed out by W. Nachtigall: Let a factory chimney stack be 140 metres high and 8 metres wide at the base. Its slenderness ratio thus

amounts to $\lambda = \dfrac{l}{d} = \dfrac{140\,\text{m}}{8\,\text{m}} = 17.5$.

Let a stalk of rye be $l = 150$ cm high and basal $d = 0.3$ cm wide. Then a slenderness of 500 is obtained by calculation.

8 Bending of a stalk of grain in wind. G is the weight of the ear, F_W and $F_{W'}$ are wind forces evoked by resistance of ear to the air.

$$F_W = \frac{\varrho}{2} \cdot v^2 \cdot c_W \cdot A$$

[ϱ = density of air,
v = velocity of wind,
c_W = resistance coefficient,
A = cross-section area of ear].
F_R is the resultant bending force.

9 A size-diameter comparison (schematic) between the Moscow Television Tower (a) and a stalk of wheat (b). The stalk of wheat was drawn in its proportions in such a way that it might withstand the loadings with such a size
($d \sim l \cdot \sqrt{l}$, and not $d \sim l$
[d = diameter and l = length of stalk]).
Of the remarkable thinness of the stalk occurring with normal size there is nothing more to be seen.

If it were to be as high as a factory chimney, "then it would need to be a mere

$$\frac{140}{500} = 0.28 \text{ m thick.}"$$

Is it thus far superior to the technical solution? The answer is no. All one needs to be protected against inordinate overestimation is to take into account the corresponding natural laws. Let us listen to what the above named author has to say on the subject: "With increasing absolute length while preserving the proportions, the ratio of cross-section area to volume in a column-shaped body gets shifted; area becomes smaller in relation to volume. The loading mass of the body is proportional to volume. Inner resistance against loading is proportional to area. Thus, with a growing increase in absolute length a point is reached when load exceeds inner resistance. The column snaps and breaks down. If this is to be prevented, then thickness d must increase in a linear way with length l (d $\sim$ l) but—according to the law formulated by BARBA-KICK ...—in proportion to the product of length and the root of length (d $\sim$ l · $\sqrt{l}$). A stalk of rye imagined as 140 m high could not therefore possess a slenderness ratio of 500, that is a basal thickness d = ... 0.28 m without breaking down. It would have to possess the following thickness:

$$d \sim \frac{1}{\lambda} \cdot l^* \cdot \sqrt{l^*} = \frac{1}{500} \cdot 93 \cdot \sqrt{93} \text{ m} = 2.7 \text{ m}$$

(l* being the quotient from the absolute length of the large-scale execution [140 m] and the original size quantity [1.5 m]). Thus the slenderness ratio of the rye stalk would now amount to

$$\lambda = \frac{l}{d} = \frac{140}{2.7} = 52,$$

thereby reaching the same order of magnitude as the chimney stack."

Despite this, flexural rigidity and buckling strength of stalks of grain or grass is an outstanding one, especially as they carry ears of exactly the same weight as themselves and, in addition, eccentrically seated ones. These ears with their awns put up a comparatively high resistance to the wind, the displacements of the stalks giving rise to heavy flexural loads. At the stalk's

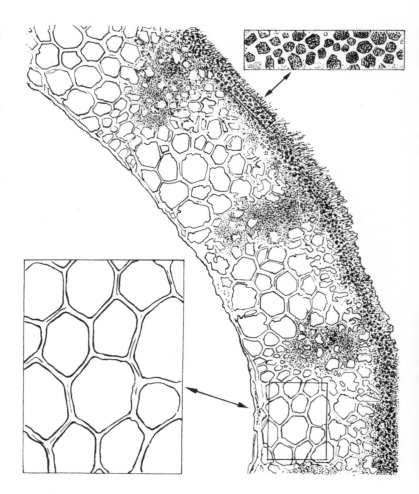

10 Enlarged sectional view of a stalk of wheat. It distinctly shows the sandwich structure consisting of inner tube and outer tube, in between the light, cellular core. Enlargements of sections show parts of the hard and the soft substances respectively. After H. Hertel.

Architecture of the Spongiosa— an Ossified Trajectorial Framework

periphery—let us imagine a cross-section—there are vertically placed sclerenchyma strips which occasionally fuse to form a radial framing. The general rule that biological structural units are inserted "in an optimal position" is best implemented here: because sclerenchyma strips are not arranged close around the longitudinal axis of the stalk but as far "on the outside" as possible, the stalk's areal moment of inertia is considerably increased. This makes the bending stability that one observes easy to understand.

The stalk wall has a multi-layer structure—the sandwich design—which particularly raises its resistance despite the ultra-light construction. This sandwich construction, in which two thin covering membranes of very high strength enclose a core of small-thickness filling medium, combines high strength with low weight. Recently sandwich panels have found their way into light-weight building; the sandwich type of building is equally known from cardboard containers where corrugated paper is often glued between two firm set pasteboard or paper covers. On all occasions, it is the light middle third that joins the two firm coverings and lends the whole panel buckling rigidity, flexural strength and firmness as to form. Since this middle third joins the coatings all along the surface these cannot buckle anywhere, or break out in any other way. While ensuring transfer of shear between the two coatings it makes for strength as well.

Of importance are still other properties of sandwich panels which, of course, have nothing to do with their strength. In view of the porous structure of the middle third heat conduction is very slight, and sound waves are also strongly damped.

It should also be mentioned that the sandwich construction is of major importance in the shafts of birds' feathers.

Who would not be overwhelmed by an elated feeling when walking amidst the rigid dignity of Gothic and the swelling plastic forms of Baroque portals? This is surely due to our sense of their aesthetic fineness, their artistic modelling and representative arrangement. However, after a while this elevated feeling may become intermingled with tacit contemplations of statical nature, and the awareness dawns on us that these pilasters, attached columns, sectional gables and presumed ornaments have also carrying functions and wherever possible to support tons of weight. Nevertheless, there is hardly anyone to whom it would occur that he has natural models for these structures before his eyes, for instance in the raw-picked tubular bones which animal tenders in zoos clear out of the cages every evening as remnants of meals devoured by the big cats.

Bones consist of a certain form of supporting and connective tissue whose basic substance is calcified. They are at once hard and elastic. Their structures are fashioned by their functions—supporting, forward motion and cavity formation. Thus long tubular and cube-shaped bones take part in the formation of extremities and joints.

If one saws a tubular bone—e. g. the brachium or thigh bone of a horse or a cow—along its length, an interesting cut surface appears. It is only in the shaft that the bone turns out to be a real tube. Its cavity is surrounded by a strong wall. Towards both ends this rind gets thinner and dissolves into the *substantia spongiosa*, a spongy but rigid bone tissue that fills the ends. The fact that the inner condition has been decisively affected by multiple loading becomes clear after a mere superficial glance. The elements of the bone are arranged in such a way that it becomes an image of a supporting column with a strong shaft and suitably designed upper and lower pressure areas. There "arise tensile and pressure lines which square up thickly the upright shaft of tubular bones and diverge apart in the epiphyses (end parts—authors' note) while they overlap in many ways." (cf. W. Nusshag, 1966) However, many bones are not merely supports but levers as well, since pressure is accompanied by tension: admittedly the bone's compressive strength exceeds its tensile strength; the former surpasses even that of granite, and this happens with the smallest material expenditure.

The characteristic "pointed arches" in the disposition of the tension trabeculae had been currently known to anatomists in the preceding centuries, though the biological importance of these constructive elements still appeared highly mysterious to them. In 1830 it was J. Herschel who stated that the spongiosa architecture is a "framework of a most peculiar design in which there occurs neither a single straight line nor any of the well known geometrical curves. However, as a whole, it does appear to be systematically arranged and is designed according to laws which remain inaccessible to our research efforts."

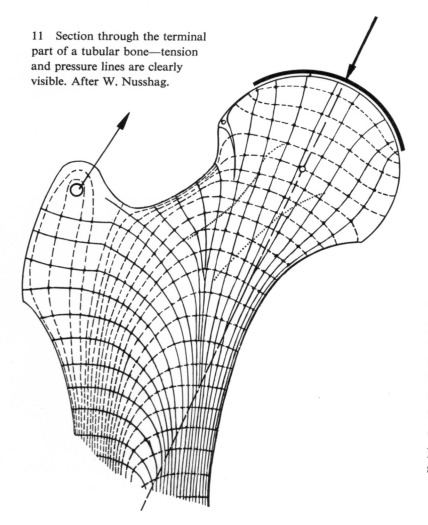

11 Section through the terminal part of a tubular bone—tension and pressure lines are clearly visible. After W. Nusshag.

In 1973 W. Nachtigall recalled K. Culmann, Zurich engineer and founder of graphical statics, who in the sixties of the last century had busied himself with designing a modern heavy-load capacity crane.

"In his work he used the method of trajectories of principal stresses. (What the expert of applied statics understands under the term of trajectories of principal stresses are curves indicating the directions of compressive or tensile stresses on each spot of the body—authors' note). In 1866 Culmann happened to enter the dissecting hall of the anatomist H. von Meyr just as he was teaching bone construction on a longitudinal section through a femur (thigh bone—authors' note). At a single glance the engineer recognized the spongiosa architecture as a system of two main tensions, cutting each other at a right angle, of an ossified field of trajectories of principal stresses. He is said to have exclaimed with enthusiasm: 'Here is my crane'!"

It was in the fifties that the principles of the spongiosa construction were conclusively explained by F. Pauwels, and in order to get acquainted with it in a little more detail we will now follow the analogy of the explanation offered by W. Nachtigall. In principle a spongiosa is built out of a couple of little trabeculae, one with compressive and the other with tensile stress, always cutting each other at a right angle. The compression trabeculae rise from the middle wall of the bone cervix steeply upward into the terminal section, the head, and end there vertically to its contour, in the area that is loaded by pressure. The tension trabeculae rise from the outer wall, cross the compression trabeculae at a right angle and end in the middle wall of the cervix vertically to the head contour. Now if the terminal section is slightly deformed by the resultant of the weight of the animal's body and the forces of the exerting muscles, then the compression trabeculae get compressed and the tension trabeculae, going vertically to them, distend. Pauwels pointed out that both systems are trajectorial frameworks which disperse in the direction of the trajectories of tensile and compressive stresses.

Let it be mentioned on this spot again that the static task set to the spongiosa is solved via construction, the spongy frame of the bone substance, with the least substrate, or material expenditure. Moreover, it now appears that the spongiosa trabeculae not only take the direction of the tensile and the compressive stresses: at the same time the thickness of the osseous substance maintains the same relation to the tension quantity in every spot. Thus the terminal section of the tubular bone is a body of equal strength and we can see that a relative minimum of substance is utilized to the best possible advantage.

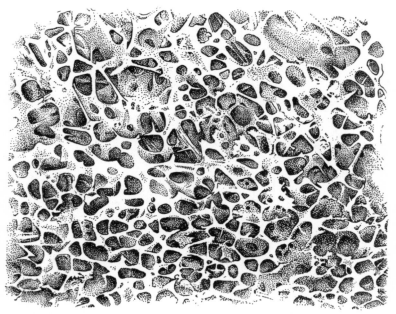

12 Section through the spongy tissue of the spongiosa, here on the femur of domestic cattle. After W. Nusshag.

Of interest is the fact that the structural aspects of the biological substrate undergo changes in keeping with the forces that actually load it. Though the spongiosa structure is no doubt genetically preprogrammed, its subtle differentiation occurs in such a way that the bone copes with those forces under a minimum expenditure of tissue. In fact, when the loading conditions get changed as a result of joint ailments or bone deformation, the spongiosa "responds" to this by an appropriate modification of its "framework". If such deviations are removed by an operation, then the framework adjusts again in a trajectory way within a few months by a reconstruction of the tissue system according to the existing load. B. Kummer (1959) wrote on this point: "During this reconstruction the individual spongiosa trabecula has to change its direction. How this happens is easy to explain by means of activity hypertrophy or inactivity atrophy if one departs from the assumption that an increase in mechanical stress within certain limits leads to an extension of the osseous substance, while a decrease in stress under a certain minimum results in its reduction."

Besides, it is not the individual tubular bone alone that is so structured that only compressive and tensile forces have to be counteracted: this principle is applied throughout the skeleton. A typical instance of this is e.g. the bending unloading of the spinal column of the neck by means of the nuchal ligament. The split-up of the bending force (the weight of the head) into tensile and compressive forces is illustrated on a rod hung up on one side by means of a joint (Fig. 13).

Another interesting aspect is comparing the relative dimensions of the bones in animals of varying sizes. It has been established that, in general, an overproportional increase in thickness is to be found in the case of large animals. This was pointed out by É. J. Slijper (1967). Such a state of affairs becomes distinctly visible if we reduce the skeleton of a lemming and that of a hippopotamus to the same size (Fig. 15).

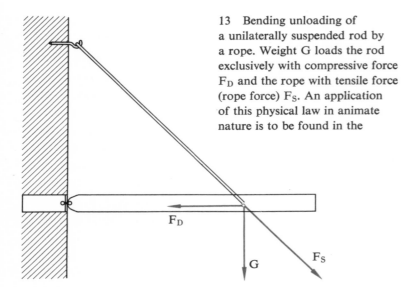

13 Bending unloading of a unilaterally suspended rod by a rope. Weight G loads the rod exclusively with compressive force F_D and the rope with tensile force (rope force) F_S. An application of this physical law in animate nature is to be found in the

F_D

G

F_S

nuchal ligament of cattle which serves bending unloading of the spinal column of the neck.

14 Nuchal ligament (1) and neck plate (2) in cattle—natural stays which ensure the carrying strength of the spine. After W. Nusshag.

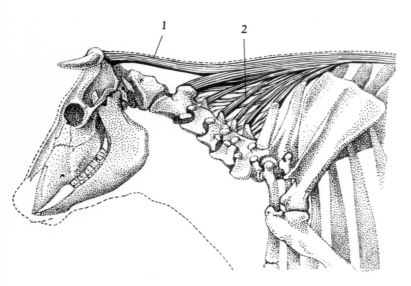

1

2

The explanation of this phenomenon has already been given above when we compared a stalk of grain with technical structures. The volume, and hence loading weight, of the animal increases with the cubic power of its size while the supporting cross-sections merely with the square (proportional increase being presupposed). If stability is not to be jeopardized, then the bone diameter must increase by $d \sim 1 \cdot \sqrt{1}$.

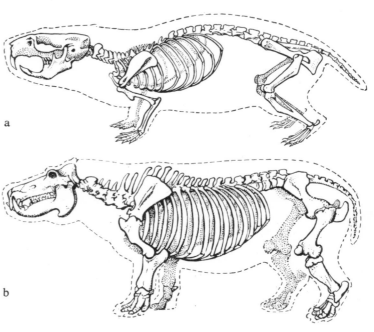

a

b

15 Skeleton of a lemming (a) and a hippopotamus (b) reduced to the same body size. The more compact osseous structure of the hippopotamus is distinctly visible. With increasing body size of animals bone diameter must grow stronger than proportionally to the length, as the body weight loading the bones grows with the third power of the size, whereas carrying capacity of bones only with the second power of its diameter. After Hesse-Doflein from E. J. Slijper.

Seeds and Fruits with Hard Shells— Natural Strengthening Elements Counteract Outside Pressure

The quintessence of all South Sea romanticism are white coral islands with slender coco trees swaying in the moist sea breeze. Nor is it often long before an atoll gets covered with these lovely palms, and the question arises how this is possible. Actually, coconuts drift on the water until they are accidentally thrown up on a beach that yields a favourable substratum for germination and further development.

Today the elegant palm of the Polynesian Islands is encountered on all tropical shores. It is interesting to note that it had not yet reached America before the continent was rediscovered in the 15th century, though its existence could be proved in those days on islands lying about 300 miles off the coast of Central America.

One-seeded stone fruits consist in the first place of a shell which envelops the seed. The shell casing is called pericarp or seed vessel. Generally this is in turn divided into an outer exocarp and an inner endocarp not infrequently consisting of but a single layer. Between the two lies the multilayer mesocarp, under it a greasy endosperm which is later opened by the seed leaf and thus made accessible to the sprout. The exocarp constitutes an effective protecting tissue (epiderm), being supported in this function by a thick cuticle (the waterproof outer lamella of the epidermis). The mesocarp is a network of fibres firmly appressed to the stone kernel (the endocarp). It is aerated and has proved its worth in the coco fruit and other hydrochorous, i.e. water-spread, representatives of tropical strand flora, as a swimming tissue and hence as a notable buoyancy device. The firm kernel is capable of withstanding multiple stress by compression—which is necessary, for how long may a coconut in certain circumstances be thrown about by the heavy surge until its odyssey ends in a quiet little place. It may be drifting for three or four months; after this, of course, it must have found its haven, as germination sets in and the embryo would perish should it find itself in unwholesome conditions of development. "About 4 to 5 years pass before the stem starts forming, and in another 6 years the first blossoms can be seen. Productiveness lasts about 30 to 40 years. There have been attempts to settle the coconut outside the coastal regions as well, but already in a slightly higher altitude where the moist sea climate is absent blossoming is disturbed.

Normally about 8 to 10 nuts ripen in one inflorescence." (S. Danert, 1973)

The trunks of these "wavering shapes" reach the height of 30 m. Should an unprotected heavy seed fall down from that height it would be sure to crack and be lost to the preservation of the species. To preclude this, nature had devised indehiscent fruits in which the seeds stay completely or partially enclosed by a pericarp during propagation. Only their outer layer retains a pulpous structure, the inner one is sclerenchymatous, built of supporting tissue and composes the stone-kernels which must be forced open by the germs.

An example of an extremely hard shell (endocarp) is the seed of the North-Brazilian palm babassu *(Orbignya martiana)* whose fruit must be crushed by force if the seed the size of a plum is to be reached. As a first-rate oil supplier this palm is indispensable to the Brazilians, and the idea offered itself of cultivating it with this purpose in view. However, its hard shell was to prove an insuperable obstacle.

A stone-fruit also familiar to Europeans and very popular for its fine flavour is borne by the *Juglans regia*, the genuine walnut tree whose recent natural habitat lies in the Balkans, in the Near and the Middle East across Iran up to Turkestan but which has been cultivated for a long time already in many regions of the Northern hemisphere. The nuts are drupaceous indehiscent fruits with a bone-hard inside layer of the pericarp —here, too, one encounters a powerful supporting tissue which offers measurable resistance to outside pressure.

This compressive strength of the brittle shells of stone-kernels and nuts is caused by the universally thick-walled stone cells or sclereids. They are part of the dead sclerenchyma, the supporting tissue in fully grown parts of the plant. On the contrary, in plant organs which are still growing one encounters only living, elastic viable collenchyma whose cell walls are thickened only in parts. According to Strasburger (1962) sclereids are spindle-shaped, often extremely protracted cells with very finely pointed ends. They have a polygonous cross-section and often a strikingly narrow lumen (inner cell space) not infrequently detectable as a mere point. Their walls are more or less lignified and correspondingly rigid.

Stone cells and sclerenchyma fibres are always so arranged that the greatest possible strength is obtained at the expense of the smallest possible material expenditure. Of decisive importance in this respect is the outer form of the fruit in question. Indeed, high tensile and compressive strength of the construction elements are not enough in themselves if the whole structure is designed in a statically unfavourable way. To serve the preservation of the species hard-shelled fruits must be capable of withstanding high impact and shock loads by which considerable forces are released. These forces should be resolved along the fruit shells in such a way that essentially only tensile and compressive forces should occur which—as mentioned above—the elements of the supporting tissue are particularly well suited to withstand. Only a shell that is polydimensionally convex can fulfil this requirement. The resolution of forces over a spherical shell is clear from Fig. 16. A force acting from above is resolved into forces parallel to the shell wall in the direction of imaginary longitudinal circles and into those parallel to equally imaginary latitudinal circles. This favourable resolution of forces lends an enormous strength to the entire fruit.

Bird's egg is another example one can quote of a relatively pressure-proof biological structure with a thin polydimensionally convex shell. A hen's egg—when taken between the palms from apex to apex—can hardly be crushed even by a strong man though the thickness of the lime shell amounts merely to about 0.4 mm.

In architecture and engineering the high strength of arches and shells has been well known and used for a long time, let us just recall the aqueducts built by the Romans, or perhaps a table-tennis ball which, despite its thin plastic skin, withstands high mechanical loads.

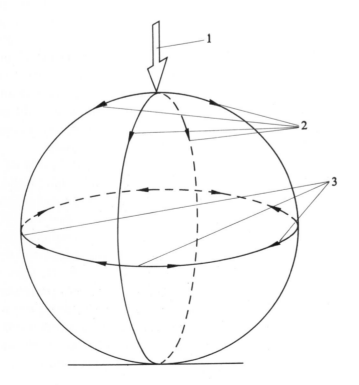

16 Resolution of a force (1) acting upon a spherical shell. Like in an arch of a vault compressive forces (2) appear along the meridians (lines of longitude). Parallel to the lines of latitude the shell is acted upon by tensile forces (3). The material shows a particularly high resistance to tensile and compressive forces.

17 A great many ears and panicles sway gently at the point of thin stalks without snapping under the load.

18 The spongy matter of the spongiosa of a tubular bone.

19 The portal of St. Mary's Church in Berlin. One can feel behind it the natural model, the architecture of the spongiosa.

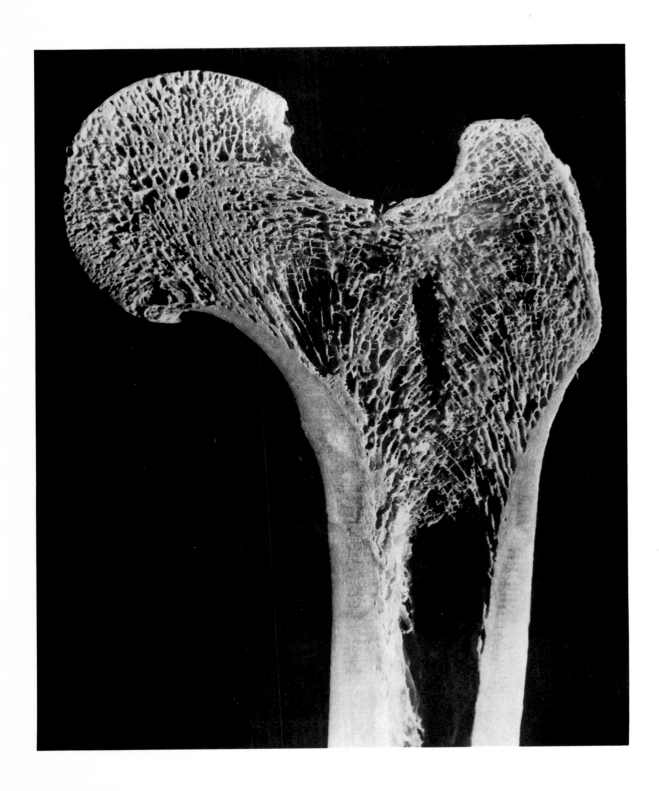

20 The bulky brachium of *Brachiosaurus brancai*, a Jurassic dinosaur from East Africa.

21 Thin tubular bones carry the heavy body of the rare African saddle-billed stork *(Ephippiorhyn-chus senegalensis)*.

22 This natural "design" is even more conspicuous in the black-winged stilt *(Himantopus himantopus)*.

23 The fruit arrangement of a coconut tree. The hard-shelled nuts symbolize the great power of resistance against pressure from outside.

24 Strong-walled stone cells bring about the compressive strength of walnut fruits.

25 The hazel-nut in European flora is a strongly diminished counterpart of the exotic coconut.

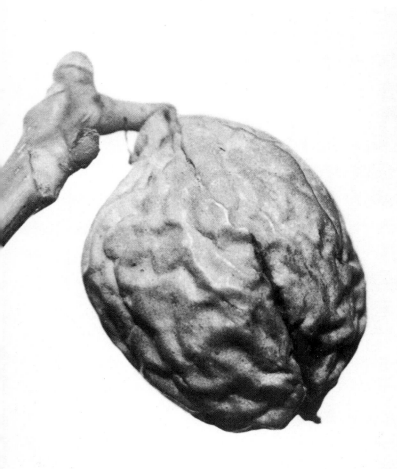

26 Sun-bathed coral reefs of tropical shores are the habitat of the emperor fish *(Pomacanthus imperator)*. It has an air-bladder.

Swimming

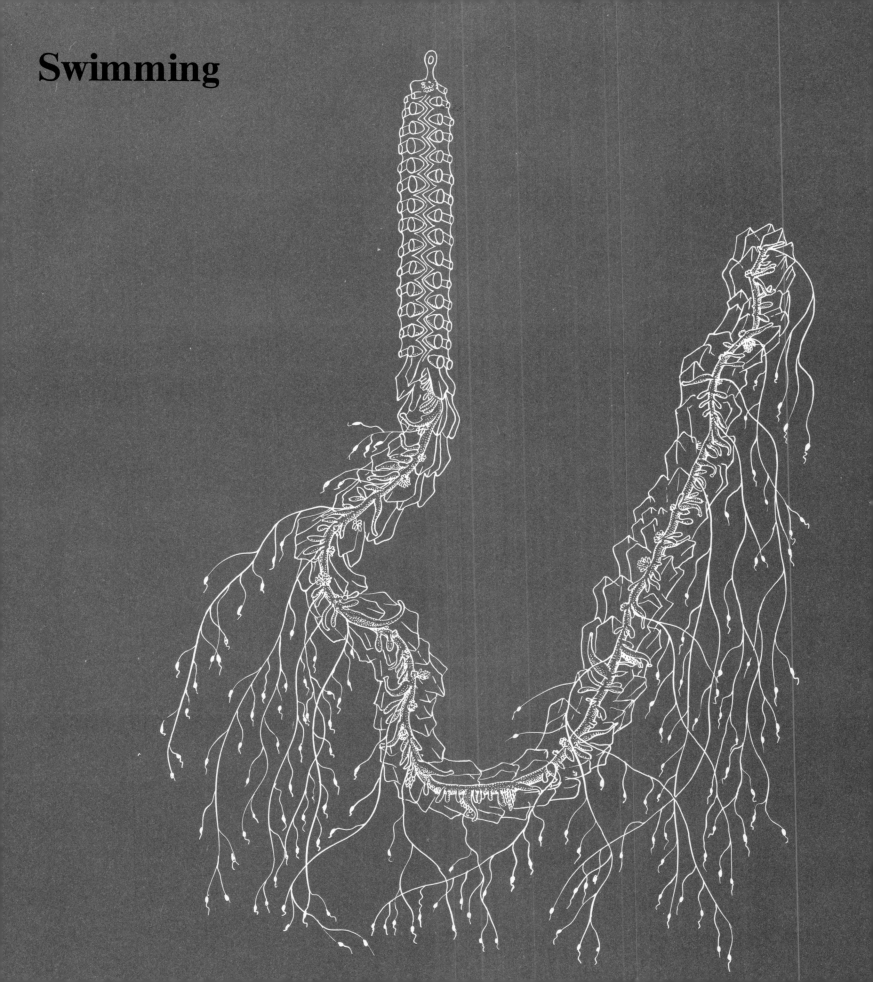

Water—a simple chemical compound of the two gaseous elements hydrogen and oxygen—and yet there is no living being that can exist on the Earth without it. Nor would life on our planet have originated as it actually did 3 thousand million years ago without the water of the primeval ocean as the necessary precondition. It was from the sea that the world of plants and the animal kingdom was finally to conquer the main land in the course of millions of years while developing an immense number of species.

Today around 71 per cent of the earth's surface are covered by oceans, littoral seas and lakes. These masses of waters contain 1,368 million cubic kilometres of water. A manifold flora and fauna animates the smallest pond provided it is not poisoned by industrial and household effluents. Animal population is to be found even in the hollows of the deep seas on whose bottom a pressure of more than 1,000 atmospheres prevails, and which is never penetrated by a sunbeam.

All water inhabitants have adapted in an outstanding way to living in their liquid environment. This applies particularly to their propulsion organs. Unicellular organisms propel their bodies by cilia or flagellum strokes, worms wriggle through the water, medusas and molluscs make use of the reaction principle after the fashion of jet propulsion, finally fishes and whales are capable of accelerating their bodies to enormous speeds by mighty blows of their tails. Many birds and mainland mammals are equally adroit swimmers and divers, often adapted exclusively for the acquisition of food in seas, lakes, or rivers.

Throughout the ages Man, too, has felt attracted to the waters. Though not equally popular at all times and in all places for personal hygiene or even for swimming, water has ever been of great importance as a traffic artery and as a source of food. By means of ships and diving apparatus Man has now pushed on to the remotest coasts and penetrated to the deepest sea bottoms.

Swimming and diving are governed by physical laws which neither technical structures nor Man and beast can escape. Some of these fundamental laws associated with the physical properties of water are to be expounded in the present chapter. Every body when immersed in water suddenly feels substantially lighter than before. According to the Archimedean principle it acquires uplift or buoyancy. This buoyancy—a force which works in the opposite direction to the force of gravity—equals the weight of the amount of liquid displaced.

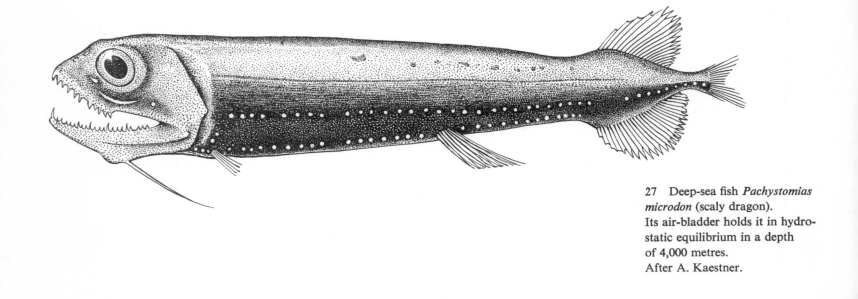

27 Deep-sea fish *Pachystomias microdon* (scaly dragon). Its air-bladder holds it in hydrostatic equilibrium in a depth of 4,000 metres. After A. Kaestner.

Archimedes (287–212 B.C.) discovered the principle of buoyancy which came to be called after him as he was engaged in a criminalistic mission on the track of an impostor. Hieron II, King of Syracuse, had commissioned him to find out whether the goldsmith had added any silver to his crown which was supposed to have been made of pure gold. The investigation had to be carried out without any damage to the crown. One day Archimedes as he entered the bath and observed the water displaced by his body overflowing the edge of the tub was struck by an idea which was to prove to be of decisive importance for the solution of the problem. The mass as well as the differing densities of gold and silver were known to him. There was but one thing lacking for determining the presence of the two metals—the volume of the crown. And the volume of this body with its irregular borders was proportional to the outflowing amount of water when the crown was immersed in a water-tub filled to the brim. Whereupon exclaiming "Eureka!" (I have found it!) he is said to have emerged from the tub and to have run naked through the streets of Syracuse. Incidentally, he at the same time discovered—having observed it on his own body—the important physical law of buoyancy. Besides, the goldsmith had actually been guilty of fraud, and was executed.

Static buoyancy, as it is also called, is of decisive significance to all swimmers and divers. Bodies of smaller mean density (mean density = weight divided by volume) than that of water swim on the surface, and immerse only so deep that the amount of water they displace reaches their own weight. They can be submerged only by applying additional external force, and they quickly rise again to the surface once the operation of that force has ceased. This is of particular importance to diving birds. If the mean density of a body is equal to that of water (1 g/cm³), then it is suspended in it and will keep in any depth without external force. This principle is perfectly utilized by fishes possessing the air-bladder. Any object lying at the bottom in the water has a density exceeding 1 g/cm³.

As we have seen, the property of a liquid decisive for causing buoyancy is its density. For quicksilver it amounts to 13.6 g/cm³ so that even solid spheres of steel (7.8 g/cm³) swim in it and become only half submerged. Static buoyancy can also be observed in gases. However, in view of the low density of such media it reaches only low values in the cases of common everyday objects. High values of air lift (air density = 0.0013 g/cm³) can thus be attained only in the case of voluminous bodies with a very low mean density such as, for instance, with those represented by hot-air balloons of the Montgolfier brothers or the hydrogen-filled air-ships of Count Zeppelin.

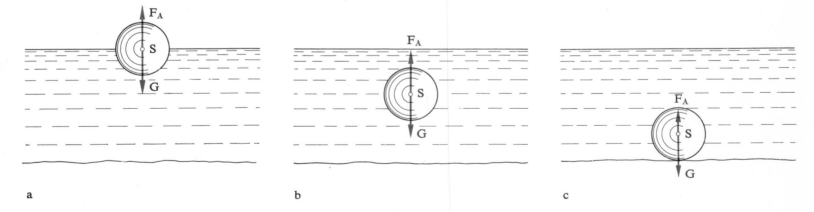

a   b   c

28 Buoyancy and swimming.
a—the body has a lower density than water, it submerges just as deep into it until the water displaced by it reaches its own weight.
b—the body has the same density as the water surrounding it (≈ 1 g/cm³), it floats.
c—the body has a higher density than water, it sinks to the bottom.

32

The density of a liquid or gas is, however, of great moment not merely for static buoyancy but likewise for drag during the body's propulsion. This braking resistance works in the opposite direction to that of the movement and is, under certain conditions, proportional to density. With velocity v it increases quadratically:

$$F_W = c_W \cdot \frac{\varrho}{2} \cdot v^2 \cdot A$$

[F_W = resistance, c_W = resistance coefficient, ϱ = density, v = velocity, A = area].

There is yet another property of a liquid that plays a major role—its viscosity. This is immediately obvious, for everyone knows from his own experience that the two liquids—water and sirup—whose density is approximately the same, offer rather a different resistance to the spoon when being stirred. Thus viscosity, shape of the body and its surface structure are all contained in the c_W resistance coefficient, which, in addition, varies with velocity.

However, the resistance offered by a liquid to a body in motion does not depend solely on the physical properties of the liquid in question, but to a high degree on its shape, or rather the profile and surface of the body as well. The resistance acquires particularly low values when the body's profile leads to a laminar flow pattern. A flow is then referred to as laminar when the respective density layers, which can be made directly visible e. g. by differing colours, glide along each other without intermixing. There is only a thin film of the liquid—the boundary layer—firmly linked with the body's surface. As a result, this film moves just as quickly as the body itself; then in the outward direction the velocity of the liquid decreases from layer to layer. The resistance is caused by friction of liquid layers against one another, and is consequently very slight.

Entirely different conditions prevail when vortices form in the boundary layer or behind the body. In this case the flow is a turbulent one. Then the body must overcome not only the friction of the fluid but also the energy of the rotating water areas, the vortices. Understandably enough, the result is largely increased resistance and therefore much higher required driving force.

Turbulent flow will appear round every body at a correspondingly high velocity even if it is streamlined. Nevertheless the critical value at which the flow turns from a laminar into a turbulent one is very strongly affected by the contours of the body and the surface roughness. Laminar profile and smooth surfaces allow of considerably higher flow speeds at a small drag than for instance unfinished angular objects. O. Reynolds discovered that on geometrically similar bodies similar flow conditions prevail provided the Reynolds number

$$Re = \frac{l \cdot v \cdot \varrho}{\eta}$$

[Re = Reynolds number, l = body diameter, η = viscosity] has the same value. This number plays a major role in all model tests with ship's hulls and airfoil profiles. Such tests with greatly reduced and cheap models of ships or aircraft yield reliable data on resistance or aerodynamic lifting force in subsequent actual craft. —If the Reynolds number exceeds a certain critical value the flow becomes turbulent. By the application of supplementary subtle natural or technical means it is occasionally possible to delay the formation of eddies to higher Re-values —and, consequently, to higher velocities. An impressive example of this is the dolphin. For a long time it had been an enigma for experts how these intelligent sea mammals were capable of achieving speeds of up to 80 km an hour, though the calculated muscle output is far from being sufficient for managing this. An explanation of this phenomenon, which has become known as "Gray's Paradox", will be given in the chapter entitled "Ideal Body Profile and Skin Structure Reduce Drag—High-Speed Swimmer Dolphin".

True Floaters in Water Medium

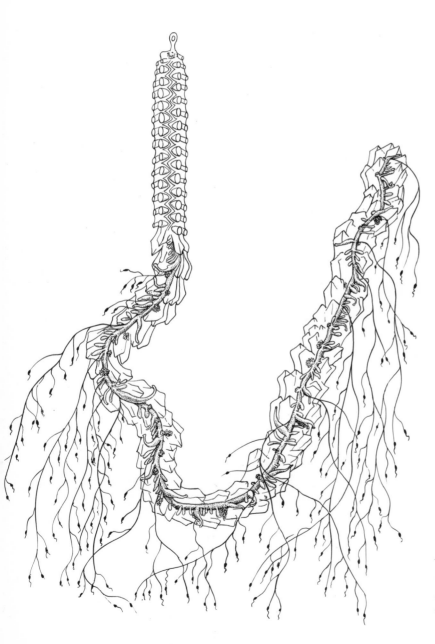

29 Siphonophore *Agalma elegans*
of the Atlantic. After A. Kaestner.

No true floaters are known to exist in air space. Considering the marvellous flying feats performed by insects, birds, or bats this may sound rather an extraordinary thing to say, but it is none the less true. Certainly, in the course of the development of the species, they have been able to make themselves noticeably "lighter", that is to make their mean density approach more or less closely that of the ambient medium. Nevertheless, they had not thereby managed to bring their bodies to floating, the latter would have required building-in some exceedingly light material in sufficient quantity. Now there are, however, hardly any creatures in our natural environment that are lighter than air. Certain gases might come into question, but then no inhabitant of the air space would fill himself with it as we do with helium or hydrogen balloons. In this respect water-inhabiting animals are better off. They comprise regular "gas floaters", and two of these, medusas with gas-filled swimming bells and fish with air-bladders, will briefly occupy our attention in the following pages.

The types of the coelenterate-order Siphonophora are called "state-building medusas" as they build colonies consisting of many individuals. As long as they live they remain linked with the common animal species and even take over specific functions in order to preserve it. As to their physical appearance we distinguish two "construction plans", which are also expressed taxonomically, i.e. in the sub-order Siphonanthae with widely extended colony forms, and the Discoanthae with a wide disk-like stem. A. Kaestner, the outstanding zoologist of the fifties and sixties, characterized the former briefly and to the point as extremely multiform marine planktons of fairy-like beauty over whose crystal-clear bodies glittering blue, red, and yellow tinges are distributed. Some of the species float and have equipped themselves at the upper end of the stem with a kind of gas-bottle, the pneumatophore. This gas-filled organ has to be imagined as a more or less large bladder of an oval to round and bulging appearance. Structurally it is to be derived from the generally known swimming bell of the medusas. "The edges close around a central gas container and, at the upper end, they still leave one pore open, or they merge with one another." (cf. in R. Kilias, 1967) At the bottom of the cavity there is a tissue of granular cells among which arise little gas bubbles that refill the reservoir which lies above them in case it has given off some of its contents through the pore. That tissue has been designated as a "gas gland".

If a group of siphonophores bearing pneumatophores is disturbed, its density changes and the medusas sink rapidly into the deep. Unless the disturbance (e.g. a mechanical irritation) is repeated, the animal will have risen to the surface again after a short time. What happened in the process? Circular muscles in the wall of the gas bottle reduced the size of the gas tank, thereby pressing some gas out of the pore; the animal stem got heavier than the medium bearing it and sank down. The gas gland took care of an early re-filling, and the tender colony regained buoyancy.

A prototype of a siphonophore with a gas-bottle is *Physalia*, the "Portuguese man-of-war", whose palm-sized, poreless gas-filled body rises above the water and does nothing but keep the animal stock afloat. There is no trace of light jelly-like swimming and floating equipments in the form of bells and leaves—as they still frequently appear apart from the Pneumatophora throughout the whole order. If the swimming body lying horizontally on the surface is struck by a breath of wind, then the whole colony sails away like a ship.

An analysis of the contents of the gas bottle container of the *Physalia* yielded a gas mixture of 15 to 20 per cent O_2, 0.5 to 13 per cent CO, N_2 and Ar approximately the same as in the atmosphere, little amount of CO_2.

Thus, with an amount of rough schematizing one can assert that siphonophores reduce the contents of their gas bottles, press these together and refill or enlarge them in order to use this active change in density in the varying depth for catching their propelling and swimming nourishment.

30 Various air-bladders, top plan view.
Left that of *Cynoscion nebulosum* (spotted sea-trout), right of *Micropogon undulatus* (California corbina). The shaded edges are tendons leading to neighbouring muscles or to body wall.
After Suworow.

31 Fishes without (shark) and with (cod) air-bladder.

This capacity is encountered once again in fishes insofar as they possess air-bladders. Air-bladders, too, usually function as hydrostatic organs that enable their bearers to float in any desired depth without sinking or rising. The density of the body is on the whole repeatedly adjusted to that of the environment. To be able to keep the animals in hydrostatic equilibrium the air-bladders must have a certain share in the volume of their bearers: in sea fishes about 5 per cent, in fresh-water fishes around 8 per cent. Salt water has a higher density than freshwater; hence the higher buoyancy known to every swimmer in sea water.

Air-bladders are descendants of the digestive tract, appendices of the foregut, which lie between the spinal column and the intestine. In many species an air passage as a link with the foregut is maintained throughout their lives. Others have this passage only in the earliest youth; as early as in a few days it gets re-formed and the air-bladder becomes fully closed.

With these physoclists the filling of the bladder with gas is provided by the "red corpuscles" in the front ventral (abdominal) bladder region, concentrated blood vessels that secrete gas. The gas secretion introduces considerable amounts of carbon dioxide, oxygen and nitrogen into the swimming organ. When a physoclistous fish wishes to give off so much gas as to obtain a normal floating condition, it accomplishes this over its "oval"—a small area of the rear upper bladder wall equipped with a thick and very fine network of blood vessels. Muscles drawing off in a radiating way from the oval edge are then contracted, the oval widens, its wall is extraordinarily thin, the gas particles wander out through it, and the many fine vessels take them up and carry them away with the blood stream. Then a muscle surrounding the oval is able to "close" it again.

Fishes with an air duct to the foregut (Physostoma) fill, or "ease" their air-bladders through this canal; they are not equipped with the structure described above.

To be sure, the change in the gas volume takes a certain time to accomplish. Secretion and absorption do not occur rapidly, and if, for instance, a fish is suddenly taken out of a great depth, the reduction in the bladder volume lags considerably behind the rapidly falling external pressure. Then it can happen that the expanding gases drive the air-bladder out of the abdominal cavity; it comes out of the fish's snout—a phenomenon well known to professional fishermen as "tympanites" or "drum disease".

Sharks and rays have no air-bladders, nor are the latter to be found in all bone fishes. They must therefore counteract sinking by continuous fin strokes unless they are anyhow in-

32 Section through a fish.
 1 Nostril,
 2 Gills,
 3 Air-bladder,
 4 Dorsal fin,
 5 Air-bladder,
 6 Caudal fin,
 7 Anal fin,
 8 Anus,
 9 Ovary,
10 Ventral fin,
11 Intestine,
12 Auricle,
13 Ventricle of the heart.
After H. Wurmbach.

habitants of the bottom whose movements are moreover limited to short strokes. Actually, a distinct correlation can be noted in general between the fish's way of life and the degree of development of the air-bladder. With most fish species in the free deep ocean the bladder is highly developed, and the gas eliminating and abducting mechanism is remarkably effective. This circumstance is of vital importance: only this enables them to participate in the extensive periodical vertical migrations of animals they feed on.

Man as a lung-breathing being is normally not equipped for a protracted stay under water. Although his medium density lies very close to 1 g/m³ he still receives enough buoyancy in water not to go down. Thus longer diving is possible only if force or additional loading is applied. Moreover, lack of oxygen, high water pressure and the consequent dangerous solubility of nitrogen in the blood prevent longer diving in greater depths (more than 10 metres) without special equipment. Therefore a corresponding apparatus had to be invented and designed, since the exploring urge, hidden riches and a great deal more drove Man already in early times to penetrate even the depths

34   Diving bell. In such a bell the diver stood up to his breast in water during his work. The air bubble compressed by the water pressure and forming in the upper part of the bell served him for breathing. Between 1663 and 1665 the Swede Albrekt von Treileben using such a diving bell and a hitherto unknown salvage process managed to recover 24 guns weighing up to two tons each

through the gun hatches of the "Wasa". The "Wasa", pride of the Swedish Navy, sank on her maiden voyage in loveliest weather on 10th August 1628. In a 32-metre depth she lay with her full equipment on the sea bottom until it was possible to salvage it in 1961.
From G. Panitzki.

33   Alexander the Great during his legendary diving attempt with a glass bell.
From H.-G. Bethge.

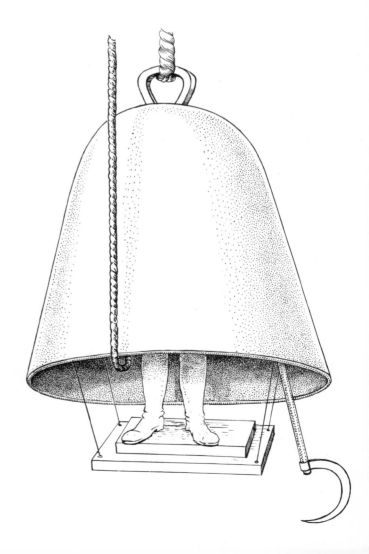

35 Eurasian aquatic spider
(*Argyroneta aquatica*) dives all
its life long. Above it is the tent
in a shape of a bell built by the
spider and filled with atmospheric air.

36 There are rather narrow limits to man's diving capacity without the use of technological devices: 30 to 40 metres and 2 to 3 minutes seem to be the record performances for depth and time respectively.

37 Technology has enabled man to go beyond the natural limits of his capability to dive.

38 A submarine swimming at high speed over the water.

of oceans. However, an independent diving apparatus or submersible boat—just like the siphonophores and fishes with air-bladders—must have a means of regulating the density and thereby also buyoancy. In this connection we wish to deal in a little more detail primarily with submarines. In the first place, however, let us mention some of the legendary diving attempts.

More than 300 years before the beginning of our era Alexander the Great is supposed to have dived in a glass bell. Almost 2,000 years later, in 1624, a wooden boat with a crew of fifteen sailed in a depth of 3 to 4 metres from Westminster to Greenwich. This year can be regarded as the hour in which the submarine was born. The first German submarine *(U-Boot)* was constructed and tested by Wilhelm Bauer (1822–1875) (cf. H.-G. Bethge, 1968). It had no ballast tanks, but during the dive (1851!) water was freely admitted for higher loading into the lower part of the boat. The existing trimming weight was not enough to keep the boat evenly balanced, and consequently water gathered more and more in the stern, penetrating moreover through leaky spots, until the boat sank. With increasing hydrostatic pressure the strength of the respective structural members was exceeded, and the fate of the "Fire-Diver" *(Brandtaucher)*, as the craft was named, was sealed; it sank to the bottom in a 15-metre depth. Fortunately, all members of the crew including the inventor managed to escape with their lives.

This setback did not deter Wilhelm Bauer or the other pioneers of the submarine the world over from trying again. As early as 1855 Bauer himself built his second *Brandtaucher* in Russia, one which made 133 underwater runs. During these underwater runs Bauer was also the first to take underwater photographs.

Analogically to air-bladders in fish, ballast tanks and the strength of the casing in relation to hydraulic pressure are of great significance to submarines. Fig. 39 shows an already thoroughly modern-looking submarine built by Holland in the USA in 1897. In the course of diving the ballast tanks get flooded with water so that eventually the ship's weight exceeds its buoyancy. If it is to be held afloat then only just so much water is let in that buoyancy and weight are in equilibrium. In general, crews in underwater craft have the atmospheric pressure maintained for them at least approximately. On the other hand, hydrostatic pressure from outside—operating uniformly in all directions—increases in a linear way with the depth, viz. by 1 atm per 10 m (1 atm $\approx$ 1 kgf/cm²). 10,000 m below the surface—in 1960 Jacques Pickard and Don Walsh with their bathyscaph named Trieste reached the depth of 10,916 metres in the Marianas Rift—overpressure amounts to 1,000 atmospheres. This entails considerable demands on both construction and material.—As to deep-sea fish, they are better off in this respect, for in their case a complete pressure balance prevails between the insides of their bodies and the surrounding water. Thus they do not feel the high pressure in any way.

39 Section drawing of a diving boat built by Holland (USA) in 1897. To be noted is the disposition of the ballast tanks whose water content determines buoyancy and positional stability of the boat.
1 Ballast tanks,
2 Battery compartments,
3 Petrol engine,
4 Electric motor,
5 Conning tower,
6 Cannons,
7 Torpedo tube.
From H.-G. Bethge.

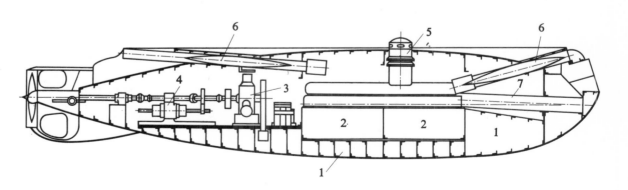

Medusas and Cuttlefishes Implement the Principle of Rocket Propulsion

More than one of our readers is sure to have had the chance to admire, in pleasantly tempered seas, large medusas in their fine transparency, and to study at leisure their ghostly-looking forward motion. One of the present authors once had the good fortune of coming across swarms of the genus *Rhizostoma pulmo* in a moderate depth on the Black Sea coast in early autumn. The first thing to amaze one was the size of these marine coelenterates, for umbrellas with a diameter of 50 centimetres and more were by no means rare. Subsequently one was attracted by the strongly developed enteric cavity which, with its multiple ramifications, the intricate cushions on the mouth arms and the lobed orifice, had once been the impetus for designating the entire order Rhizostomae, i.e. medusas with ramified mouth.

The high-vaulted jelly-like umbrellas by their contraction in a periodic, pretty slow rhythm evoked even more strongly the impression as though the diving beholder were moving amidst a procession of pale ghostly figures. Even so the regular pulsations of the bells allowed these marine invertebrates to steer their course in a "stubborn" enough way; when one wished to avoid collisions one had to hit the vertex of the umbrella with one's palm. As it were "helpless", the medusa then lingered for a few seconds, at the same time sinking a little, and finally altered its swimming course. It was obviously the oral pipe, pretty heavy as a result of its manifold appendices, that offered the water a certain resistance. During every swimming stroke it was dragged lengthwise and followed the umbrella only after a slight interval.

The principle according to which these animals move is the rebound: the medusas while pressing water out of their bell are themselves propelled in the opposite direction. In a way, they can be said to be pushing themselves off from the water they have forced out.

The physical basis on which this form of propulsion movement rests is the universally valid law of mechanics; that of reverse action formulated by Isaac Newton (1643–1727): to every action there is an equal and opposite reaction ("actio = reactio"). The quotient from the impulse of the water forced out (impulse = mass times velocity) and of time within which the water flows from the bell is one of these two opposing forces. According to the law of reaction, the thrust acting in the opposite direction and propelling the medusa must have the same value. Therefore they can be brought together by using the equality sign:

$$F_V = \frac{m_W \cdot v_W}{t_A}$$

[F_V = thrust, m_W = mass of the water forced out, v_W = its final velocity and t_A = time of ejection]. It can be deduced from this simple formula that with shortening contraction time and likewise with increasing velocity of the ejected water jet thrust, and consequently the medusa's speed, grow. To achieve this, the bell must be capable of contracting in a pulsating way.

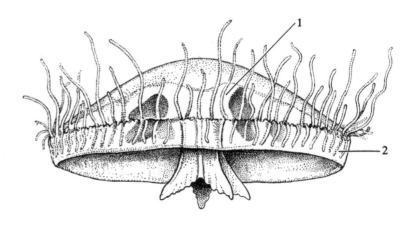

40 Bell (1) and velum (2) of the fresh-water medusa *Craspedacusta sowerbyi*.
After Reisinger from A. Kaestner.

This is possible by a cyclically arranged muscular system in its bottom wall; with many species even more so by the velum (cf. Fig. 40), which according to A. Kaestner (1965) is "annular stop-like, mostly horizontal, muscular duplicative texture" fringing the umbrella edge. It has the capacity of even more narrowing down the umbrella hollow, thereby enhancing the velocity of the water driven out during the umbrella pulsation, and increases the medusa's speed. The elastic jelly of the bell operates as the antagonist of the contracting sphincter muscles; when these slacken the umbrella also expands again and

the arising suction causes its hollow to fill with water once again.

Several pulsations are always followed by a brief pause. During this the animal always sinks, of course only little, for sinking is considerably opposed by the parachute-like form of the body and the low density (the jelly portion in the volume being high). In addition to this, the liquid of which the jelly for the most part consists is lighter than the medium that bears the medusa.

A "conducting" network system of a number of peripheral nerves spontaneously stimulates the swimming muscle system, thus causing the "automatonlike" rhythm of its contractions. Their regularity is what distinguishes the movement of medusas from a similar one of the cuttlefish, marine cephalopods. Here we once again encounter the principle of rocket propulsion. However, in this group of invertebrates it is used to accomplish also entirely different performances. These can hardly be described in a more graphic way than with the words of Thor

41 Multiplicity of forms
in coelenterates.
1 *Aurelia aurita,*
2 *Cyanea lamarcki,*
3 *Rhizostoma octopus,*
4 *Chrysaora hysoscella,*
5 *Cotylorhiza tuberculata.*
After Kosch-Frieling-Janus
and Eigener.

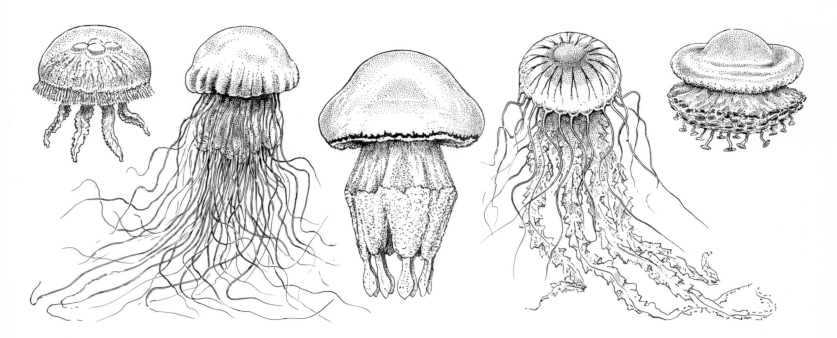

Heyerdahl, the famous raft sailor of the "Kon-Tiki" Expedition: "... We had already got three different species of these animals on board. But as the unknown strangers were approaching and some individuals were sailing in the height of 1.5 m above the raft one of them thrust straight in front of Bengt's chest and fell on board with a smack. It was a young cuttlefish. We were greatly astonished. It has been known for a long time that the cuttlefish swims on the rocket principle. It pumps sea water with great force through an open pipe on the side of the body, and is able to shoot the body backward by jerks in a whizzing spurt, and when it links all its tentacles in a thick bundle above its head it becomes streamlined like a fish. At the sides it has two fleshy skin folds which it generally uses for steering and slow swimming in water. However, it also turned out that irresponsible cuttlefish youths, the favourite dish of many large fishes, managed to escape from their pursuers by driving through the air in the manner of flying fish.

They had put into practice the principle of rocket flying long before human genius hit upon this idea. They were pumping sea water through their bodies until they were tearing along, and then they steered their course diagonally upwards through the water surface stretching their skin folds like wings. After the manner of the flying fish they thus sailed in a gliding flight as far as their impetus would carry them. Since we became aware of them we often saw them sail 40 to 50 m through the air, singly or in flocks of two to four."

42 Multiplicity of forms in molluscs.
1 *Pyrotheutis margaritifera* (ventral view),
2 *Abralia veranyi* (dorsal view),
3 *Sepia officinalis* (ventral view),
4 *Gonatus fabricii* (ventral view),
5 *Stenoteuthis bartrami* (dorsal view).
After Kükenthal/Matthes.

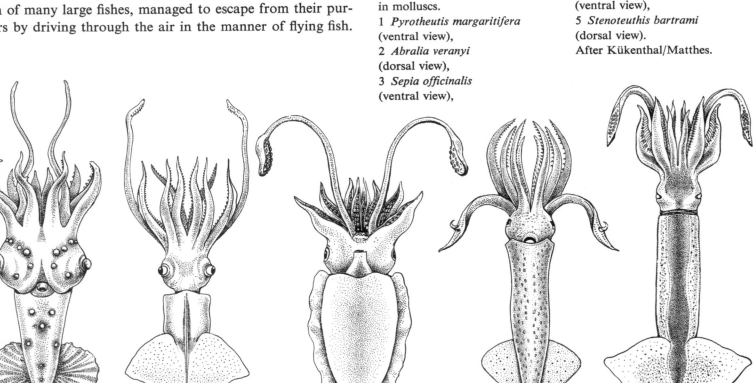

1   2   3   4   5

If the reader is to understand this fully, a few words on the overall anatomy are necessary. The trunks of cephalopods are compressed into the form of bags, or have a slim torpedo shape. Between head and trunk there is a funnellike organ—a movable tube linking the shell cavity with the water outside. It functions as a "return stroke" for breathing water which flows through an aperture on the head into the shell cavity. When the strong muscular system of the shell wall contracts the cavity, the inlet closes, and thus the only way left for the water to come out is through the narrow funnel tube. This "jet" is capable of passing through it with considerable speed, and since the funnel can be rapidly bent in most diverse directions,

it is possible to alter the direction of swimming abruptly. The recoil brings about regular rocket launchings; according to eye-witness accounts sepias and loligos vanished from their field of view as though shot out of a pistol. Fin edges or fins and arms (tentacles) are used for steering and stabilizing purposes. Fig. 43 illustrates very clearly that the flying loligos (genus *Stenotheutis*) combine rocket launching and gliding with outstretched fins and membranous edges of the outer tentacle as bearing surfaces. They describe a curved trajectory for which heights of 2.5 metres and lengths of about 15 metres were established by F. Arata in 1954.

Among the active flyers of the animal kingdom, whose life element is the air, rocket propulsion is something one looks for in vain. This, however, is quite easy to understand since for acquiring appreciable thrusts—in view of the small density (and thus also mass) of the outflowing air—very high ejection speeds would have to be reached. What speaks unequivocally against jet propulsion in the biological sphere are enormous anatomic problems and a low degree of efficiency at relatively low flying speeds. In this element it is much more favourable to move forward by wing-strokes whether in gliding or sailing.

43   Take-off and flight of flying loligos. The water driven out of the funnel (left below) allows these cuttlefishes to leave their habitual medium like a rocket. Swimming edges (right) spread as supporting surfaces make a true gliding flight possible.
After W. Klausewitz.

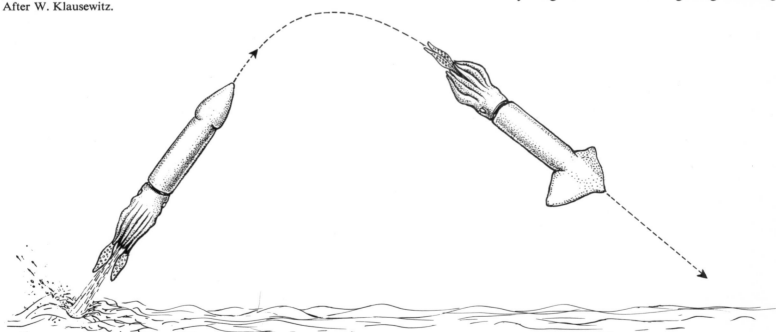

In engineering, however, the position is quite the reverse. There jet propulsion has gained a dominant place in flying and in space-travel rockets. The fact is that since the early thirties flying engineers have ever more come to realize that propeller drive is not suited for higher speeds (above 700 to 800 km/h). During their precipitate revolutions propeller tips moving against the onflowing air locally develop sonic speeds with the resultant shock pressure and a corresponding drop in power. The rocket known for centuries and allegedly to have been invented by the Chinese turned the attention of designers to the possibility of propelling aircraft by a highly heated gas jet on the reaction principle. Tests with gunpowder rockets attached to the aircraft brought no satisfactory results as there was no way of controlling the explosion. Taking up the idea of the gas turbine then known already for several decades aircraft designers, particularly in Germany, England and also in the Soviet Union, were working in the thirties with a view to developing jet propulsion units. The world's first jet aircraft, the "He 178", flew on 27th August, 1939 powered by a jet propulsion unit developed by Pabst von Ohain at Heinkel's in Rostock.

After the Second World War jet propulsion rapidly rose to importance in civil aviation as well. One of the best known and safest air liners dating from the fifties has been the Soviet "TU 104". Nowadays transcontinental air traffic is difficult to imagine without jet planes.

The way a jet propulsion unit operates is illustrated by our schematic drawing (Fig. 44). The air streaming in from the left is compressed by a multilevel compressor and driven under pressure into the combustion chambers. There the injected fuel —this can be one of low quality—burns in the oxygen-nitrogen mixture at a high temperature (about 800 °C). The hot combustion gases escape at a high speed (over 500 m/s) from the thrust nozzle but prior to that they drive a turbine wheel mounted on a shaft along with the compressor.

44　The principle of jet propulsion.
a—forward motion (2) of a medusa by ejected water (1).

b—diagram of a jet propulsion unit:
1 Air compressor,
2 Combustion chamber,
3 Gas turbine,
4 Thrust nozzle.
Partly after W. Just.

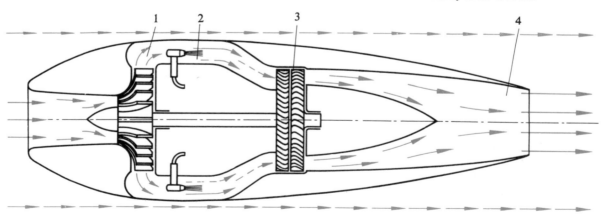

a　　　b

The efficiency of a jet propulsion unit is determined by the thrust, which amounts to several thousand kilogramme weight in the case of modern equipment. The thrust can be equally established by using the simplified equation for medusa's propulsion, naturally with due account being taken of the fact that the hot gases stream out continually and not intermittently like the water out of the bell. The mass of the combustion gases grows with time:

$$m_G = \varrho_G \cdot v_G \cdot A_s \cdot t$$

[m_G = gas mass, ϱ_G = its density, v_G = speed of flow, A_s = jet sectional area, t = time].

By inserting this relationship into the "medusa formula" the following equation is obtained: $F_v = \varrho_G \cdot v^2{}_G \cdot A_s$, an equation that can be applied to the lifting jet of hummingbirds (see chapter "The Hummingbird—a Miniature Helicopter").

Rockets are also propelled by a hot jet of gases issuing from nozzles. Of course, its propulsion units do not take the oxygen necessary for combustion from the air but carry it along chemically bound or liquefied. As the rocket acquires repulsion from its gas jet it can also fly in a vacuum, the vacuum of the Universe. High-powered rockets had been the prerequisite for manned space travel which was inaugurated by Yuri Gagarin first orbiting the Earth in 1961. In 1969, hardly 12 years after the launching of the first earth satellite, "Sputnik I", Neil Armstrong and Edwin Aldrin were the first men who set foot on another celestial body—the Moon.

The following quotation from a publication by Prof. Ernst Schmidt, formerly employed at the *Deutsche Forschungsanstalt für Luftfahrt* (German Research Institute of Aviation) shows that as late as 1954 not even experts could foresee such a tempestuous development of rocket engineering and space travel: "As for reaching the Moon, or orbiting it, in manned rockets, I do not consider this technically impossible but, in view of the monstrous expenditure, impracticable. On the other hand, travelling to the Mars must, in my opinion, be regarded as a utopia from the technical point of view." The fact that even an—unmanned—rocket-propelled journey to Mars is no longer a utopia was demonstrated by the landing module of the U.S. Viking 1 Mars probe which made a soft landing on the "Red Planet"—about 320 million kilometres away from the Earth—on 20th July, 1976. The probe had been launched on 20th August, 1975.

Ideal Body Profile
and Skin Structure Reduce Drag –
High-Speed Swimmer Dolphin

"Nowadays any schoolboy will tell us that whales and dolphins are no fish but mammals, and as early as in the year 400 B.C. the great Aristotle tells us that the Cetacea possess hair, that they breathe not with gills but by means of lungs, are nurtured after birth with mother's milk, and that the animals' caudal fins are horizontal and not vertical as in the case of fish or reptiles. The fact that in spite of this not only Aristotle but also Pliny (about the turn of the era), BELON (1553) and RONDELET (1554) had placed these animals with the fishes in their systematic classification is to be ascribed to the circumstance that what these authors used as the basis for their classification of the animal kingdom was the environment where the animals live. It was JOHN RAY who was the first to classify the Cetacea with the mammals in 1693 and in LINNAEUS one can already find the subdivision into the toothed whales (Odontoceti) with teeth in their jaws, and the toothless or whalebone whales (Mystacoceti), which have no teeth but capture their nourishment with the aid of baleens ...

The general shape of the body of whales is that of a fish or torpedo, a streamlined form, i.e. one lending it the smallest possible resistance while swimming in water. The skin is smooth, and only near the jaws are hairs, or rudiments of hairs, bearing the character of sense hairs. In order to ensure a regular laminar flow of the water along the body practically all organs protruding from the body in other mammals are built into this streamlined shape. There are no external ears, no protruding teats, and even the exterior parts of the hind limbs are completely absent ... The fore limbs are ... always grown out into flat, more or less long paddles which still possess all the skeleton elements of a normal foreleg and just like the hindlegs are constructed in young embryos like normal mammal limbs.

This points to the fact that the Cetacea are descended from normal mammals living on the land ...

The lives of whales and dolphins take place mainly in the obscurity of the water. Even so the traveller is apt to see more of these animals than of fish as the former are compelled to appear on the surface in order to breathe. Thus dolphins accompanying the ship in front of its bow are a familiar sight to every seaman and now and then he may even encounter big whales."—

We may be blamed by the reader for the length of the quotation. But we would not know any better and more concise general characterization than the above given in the words of E.J.Slijper, the Dutch zoologist, probably the best specialist on the Cetacea in our days.

When whales—both big and small—are swimming in the normal way, then neither pectoral fins nor dorsal fins have any part to play in it; what they provide is more or less the stabilizing and steering of the thick spindles of the body. The forward movement in the wet element occurs by the movement of the tail with the horizontally set "fluke", the caudal fin.

Film shots of swimming dolphins have shown that the to-and-fro movement of the tail is mostly carried out in the regions of the tail root, i.e. where the anus is situated. Otherwise it takes place in the transition area of the end of the trunk into the caudal fin. Thus the force exerted by water resistance during the fluke stroke is directed diagonally top front on impact and down front on rebound. Both forces can be divided into two components, one of which is directed to the fore and one, perpendicular to it, pointing upwards and downwards respectively. The latter two force components cancel each other out, while the forward pointing components lend the dolphin propulsion during both impact and rebound. "Anybody who cannot very well imagine," we should like to add with Slijper, "that such a small surface as the caudal fin can propel the whole animal body, should have a look at a ship in a dry dock and see how small the propellers of a mighty ocean steamer really are." It is little known that whales and dolphins combine three stroke effects or vibrations: the vertical stroke (a specific utilization of the caudal fin), strokes into both sides (here the high side-surface of the hind parts is used) and twisting the tail round the longitudinal axis.

Many whales, large and small, have long been known as regular swimming champions though a distinction must be made between "sprinters" (with only brief spells of the highest speed) and the "marathon runners" (swimming at normal speeds for several hours). Let us illustrate this by giving a few examples: large rorquals usually travel at a speed of 12 to 14 knots (1 knot = 1 nautical mile per hour = 1.85185 kilometres per hour). For fin whales and *Balaenoptera borealis*, which belong

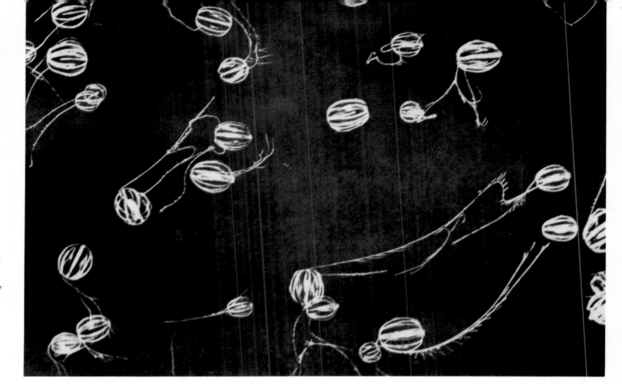

45 Reaction propulsion like the one encountered in jelly-fish is impossible in the case of the crystal-clear comb jelly. These partly row with the help of cilias arranged in riblike fashion, partly let themselves be borne by ocean currents.

46 *Chrysaora hysoscella* swims by means of quick pulsations of its bell.

47 The recoil of hot gas jets drives the rocket weighing hundreds of tons from the take-off platform within a matter of seconds.

48 Take-off of a Polaris-A-2-rocket from the atomic submarine "Henry Clay" (USA) weighing 7,250 tons. The pieces flying off the starting rocket are parts of starting facilities.

49 A living rocket: a cuttlefish.
Here the reaction apparatus, the
funnel organ, is hidden between
the mouth arms.

to their category, 35 knots (65 km/h) have been established as their top-level performance. Right, Californian grey and humpback whales manage 18 km/h at the most, normally about 5 km/h. Sperm whales with their 18 to 40 km/h are considerably faster, though they can hardly be compared with their gigantic colleagues. Dolphins and porpoises normally manage to make 40 km/h, which is surprising as they are even considerably faster in sprint (H.-G. Petzold quotes 80 km/h). However, there is a peculiar phenomenon to be noted in these small elegant toothed whales, i.e. that they manage to keep step, just as their gigantic cousins by comparison, with modern transoceanic steamers—"the most remarkable thing about the whole matter is, however, that the speed of the small porpoises and dolphins is exactly the same as that of their relatives whose size (volume or weight) is about a thousandfold of their own. Indeed, it is widely known about ships that in general speed increases with the tonnage." (Slijper) The astonishingly high speeds of the small whale species are not understandable if normal streaming conditions along their bodies are assumed. In spite of the streamlined shape of a dolphin's body resistance would still assume such high values that the existing (calculated) muscle performance would be far from sufficient for overcoming it. There are two possible explanations of this phenomenon known as "Gray's Paradox": Either the specific muscular performance of dolphins widely exceeds that of all other mammals —or else the drag is so reduced by a "trick" that normal muscular performance is sufficient to ensure the blazing forward motion. Actually, however, this is not a case of the specific muscular performance being abnormally high but rather the drag is reduced by natural "devices" to about one-tenth of the value theoretically calculated for the dolphin body. In this respect the dolphin is far in advance of technology. In terms of fluid dynamics a small drag is of equal importance as full laminar flow past an object—a flow without turbulence in the boundary layer and behind the body. As long as the flow remains laminar, drag increases only in proportion to velocity (Stokes Law). When turbulences occur, then drag increases abruptly and rises with a square of velocity.

The phenomenon of the extremely small drag is not to be explained only by the laminar form of the dolphin body in which, as described in the introductory quotation, all "projections" are smoothed. The fact is that with sufficiently high velocities (high Reynolds number) flow on rigid, smooth laminar bodies also becomes turbulent as the pressure at the stern increases, which leads to eddy formation by back-surge. To obviate this phenomenon in flying suction of the boundary layer has been successfully employed. It was Ludwig Prandtl (1875–1953) who discovered the importance of the boundary layer for flow processes as early as 1903 and proposed the suction of the boundary layer in 1904.

50   The dolphin with its most important body characteristics.
1 Dorsal fin,
2 Caudal fin (fluke), which in contrast to caudal fins of fish is set horizontally,
3 Anus,
4 Lacteal glands,
5 Sexual opening,
6 Pectoral fin,
7 Auricular opening,
8 Eye,
9 Snout,
10 Front of head,
11 Nostril.
The streamlined shape is practically undisturbed by the body organs. After H.-G. Petzold.

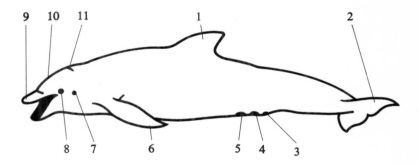

However, what possibilities has the living creature dolphin to maintain its boundary layer laminar?—For one thing, there is the complex fluke stroke by means of which pressure conditions at the stern can be made more favourable. None the less, neither the ideally adapted body shape nor its movement are enough to reduce the drag to such an extreme degree. Here the most significant part is played by the epidermis on whose boundary with the water the decisive flow processes occur. It vibrates with the boundary layer wave that arises with the starting turbulence and damps it. By this means those dangerous early signs of turbulence are subdued. The epidermis, which is only about 1.5 millimetres thick, is best designed for this specific task. It consists of a 0.5 millimetre strong, folding membrane and of an absorbing layer underneath it, 1 millimetre thick and elastically formable, consisting of as many as 80 per cent of water. This water is able to flow parallel to the surface in finest little canals. Underneath lies the inner skin, about 6 millimetres strong, composed of a very firm and viscous tissue. Vibrations in the boundary layer are transferred by the soft outer skin to the damping layer which, when deformed, causes the water to flow through the narrow little channels into the other areas. The tiny channel diameters brake the inner flow in this layer, thus damping the vibrations much the same as in a motor-car shock absorber. All these are passive processes which take place automatically during corresponding flow conditions and are not influenced by the dolphin's nervous system.

Similar conditions of flow occur on similar bodies only when the Reynolds number

$$Re = \frac{l \cdot v \cdot \varrho}{\eta}$$

acquires the same value (see introductory chapter). Therefore, the greater lengths and the form of the Balaenopteridae (rorquals) make a partially turbulent flow probable with these animals, since with the larger diameter l the Reynolds number does not increase only then, and thus may cause turbulence, if velocity is lower. Possibly because of this they are not faster than their much smaller relations.

Here, of course, is the added fact that the "energy-processing machine" (H. Hertel) in porpoises and dolphins, i.e. their dorsal and caudal muscular systems, is much more effectively fixed to the long spinous process in the caudal region than in the very large representatives of the family. The revertive strong and long muscles are attached to the rear transverse processes of the lumbar and caudal vertebras (cf. in W. Jacobs, 1938).

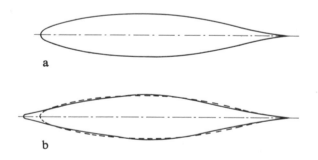

51 Comparison between the NACA 66018 laminar airfoil profile (a) and a dolphin (b). For better comparison the NACA profile is once more drawn into the outline of the dolphin's body with a dotted line. After H. Hertel.

Some anatomic peculiarities enable the dolphin's muscles to operate with far greater lever arm than that of a large whale. That this kind of combination of functions in the bodies of the top-level swimmers in the animal kingdom yielded excellent examples and gave manifold incentives to designers is, incidentally, easy to understand.

A strongly reduced drag would lead to enormous saving of fuel in navigation. Attempts in this direction have been, and are still being, made the world over. As early as 1938 the German W. O. Kramer took out a patent aimed at reducing drag.

Turbulence formation was to be repressed by wires stretched in the direction of the flow. A similar role is ascribed to certain feathers on the leading edge of birds' wings, and the owl's noiseless flight is to be explained by a turbulence-free flow past its wings. After the Second World War W. O. Kramer carried out extensive researches in the USA with dolphin skins, and designed an artificial two-layer damping skin similar to that of the dolphins, using rubber and silicon preparations. By means of this skin the drag of test profiles could be reduced to about 40 per cent.

Presently intensive experiments with dolphins are being conducted all over the world. Though the priority object of such experiments is to explore the behaviour of this highly intelligent sea mammal, it is nevertheless certain that interesting results can be expected for flow technique as well.

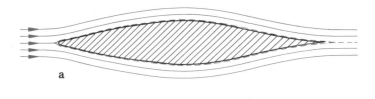

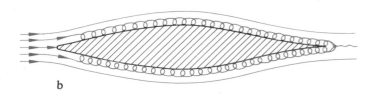

52 Flow boundary layers at a profile.
a—laminar boundary layer; the liquid layers glide along one another without getting mixed. Drag is small.
b—turbulent boundary layer in which the liquid layers are mixed together into a whirl. Drag is larger than with laminar flow.

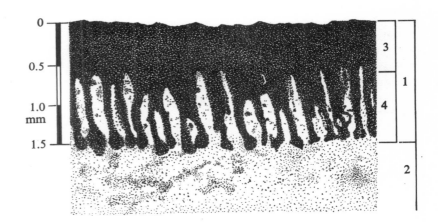

53 Section through the flow-damping dolphin skin.
1 Soft outer skin,
2 Firm inner skin,
3 Outer membrane,
4 Damping layer.
After H. Hertel.

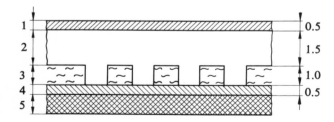

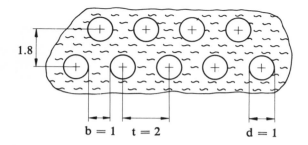

b = 1   t = 2   d = 1

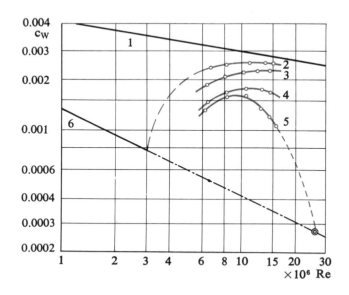

54   Technical imitation of the dolphin skin.
1 Outer skin,
2 Soft layer (rubber),
3 Liquid (oil),
4 Inner skin,
5 Body
(all dimensions in millimetres). Turbulences setting in at high speeds set the soft layer and the liquid through the outer skin into oscillations.

Thanks to the high drag in the canal-filled skin, its oscillations and along with them the turbulences in the outer fluid flow are damped and suppressed. The drag of the whole body is strongly reduced in this manner. After W. O. Kramer.

55   Drags of technical dolphin skins. Plotted is the resistance coefficient c_W above Reynolds number Re, which contains velocity.
1 Fully turbulent flow,
6 Fully laminar flow (both theoretical),
2, 3, 4, 5 different artificial dolphin skins which strongly reduce the resistance of a rigid body of the same shape.
After W. O. Kramer
from H. Hertel.

Constructive Adaptation to Aquatic Way of Life—Diving Birds and Mammals

Many vertebrates have resorted to water, the primeval living space, in order to gain from it one of the most important preconditions of their existence—their food. It is abundantly clear that this involved a multiplicity of constructive adaptations in anatomical and physiological respects, even though the divers were not completely alienated from land, their original environment. Nowadays birds and mammals hold their own among a distinguished number of diving representatives, and should one wish to deal with all of them, then our limited literary framework would soon be blasted. Therefore we shall do with a glance at penguins, divers and seals; they themselves present so many interesting diving aspects that one finds it hard to make a brief choice.

In the first place: diving is a question of density. The diving body should be as heavy as the medium in which it moves. Subsequently the difference between a bird's as well as a mammal's body and the water becomes strongly obliterated; all the better can the feeding space be mastered. Penguins and divers are introduced by two adaptation forms perfect in this particular. Quite contrary to the bones of birds which are preferably at home in another element, their bones are not pneumatized, thus they are free from air—easy to understand, since they needed the "weighting" of their matter in order to counteract buoyancy.

Another means to this end consists in decreasing the air content in the plumage, since the air contained in the feathers actually endows the bird under water with a stronger uplift.

Therefore the plumage undergoes a transformation; it lies down flat along the body and thus produces a good sliding surface. Feathers are separated into rough fibres resembling hair and in the case of penguins cover the whole skin. The unusual way of life of penguins which inhabit the Southern hemisphere in 17 species has a marked bearing on their outward appearance. Though at times they move rather clumsily on land, they are extremely dexterous and fast under water when catching fish, crabs or swimming molluscs. Their spindle-shaped bodies favourably built from the point of view of flow technique are propelled by wings, the webbed feet taking over the functions of steering and braking.

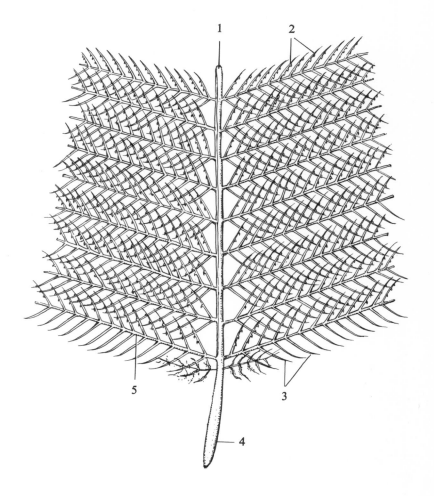

56 Schematized structure of a contour feather on a bird's body.
1 Shaft,
2 Barbs,
3 Barbules,
4 Quill,
5 Aftershaft.
After H. Frieling from R. Berndt and W. Meise.

Penguin wings are narrow, flattened in sickle fashion and comparable to fins or oar blades. Pinions which would enhance water resistance are no longer in existence; their surface is covered with thousands of scaly smooth tiny feathers. No wonder then that these birds of faraway latitudes can as it were fly through the water—a recent film from the Antarctic by J. Cousteau has given outstanding evidence of this capacity. The wings strike from forward above to backward below while moving them by rotating round the longitudinal axis with the narrow edges in the forward stroke but with the sliding surfaces in the backward stroke.

Strong muscles attached to the breastbone crest and to the broad shoulder-blade enable the animal to make a considerable number of strokes and at great speeds. In the king penguin *(Aptenodytes patagonica)* about 120 wing-beats a minute have been counted, the smaller species reaching as many as 200 strokes. The fastest penguins make about 35 km/h.

There is yet another form of adaptation that can be exemplified in loons and grebes (families Gaviidae and Podicipedidae). Their bodies have also been remodelled by evolution into torpedo-like narrow ones, and thereby made outstandingly capable of diving. The wings, of course, still serve for moving in the air, for particularly the Gaviidae are hardly able to withstand the rigours of winter in their Northern brooding places—they have to escape to the open sea and to larger inland lakes in southern plains. Even the Podicipedidae which are spread throughout the world migrate in the cold season to warmer

57 Birds diving. Above: grebe,
left: snakebird, right: penguin.

climates. Grebes carry broad lobes of skin on their feet; loons, on the other hand, are equipped with ducklike webs.

In both cases their art of diving is impressive though on land the birds are almost touchingly helpless. Thus for instance many of them cannot take wing from land, and even from water only after a longer run-up. On the other hand, they withstand diving depths up to 50 metres and diving times of 10 minutes.

Swimming is done by the legs; propulsion takes place far behind, and under water both webbed surfaces work simultaneously. According to W. Jacobs the force generated by the leg strokes issues from the part of the body where legs are anchored, that is in the first place in the hip-joint which lies right at the back on the level of the vertebral column. This is of special importance to the diver as the front part of the body is lighter than the hind part burdened with mighty leg muscles; the focus thus lies rather far behind. "Actually, the bird should therefore find it difficult to get into the depth; the front part of the body would constantly tend to straighten up. However, in view of the rear position of the hip-joint the body makes a tipping movement with each of the strokes following each other in quick succession. Moreover, this steering into the depth is substantially promoted by the peculiar way the legs are held during the swimming stroke. Indeed, at the moment of the stroke the legs are straddled far outward and turned upward in such a way that the toes look nearly upward. Tarsus and toes when seen from the side appear far above the back." The lower segment of the leg would thus appear to be rolled outwards by about 90 degrees in the knee-joint, only then primarily the joint between the lower segment of the leg and the tarsus is vehemently stretched. Thereupon the lower leg is turned back and the leg is brought forward again before the next stroke.

This remarkable twisting movement is made possible by a peculiarity in the skeleton. "The articular eminence and the cavity of the joint on the knee are placed in such a way that plentiful revolving movement can take place." All diving birds have a high pivot on the shin. It protrudes considerably above the knee-joint in a forward or upward direction. In loons it is nothing more than a continuation of the shin. In grebes the knee-cap is firmly grown to the shin and very much enlarged. That pivot is the starting point for various muscles which stretch the knee-joint, turn the shin upwards, stretch the tarsus for the stroke, etc.—Jacobs was fully justified in stressing the fact that the remarkable use of the knee-cap in the service of diving is an entirely unique phenomenon.

Quite a different way of diving—thrust diving—is practised by some pelicans and gannets but also by the sea-swallow and the kingfisher. This was reported by H. Tributsch in 1974.

These birds are lightly built and despite heavy lift dive 15 to 20 metres deep. As we know a body whose lift is higher than its weight can dive only when there is some additional outer force working in the direction of weight. And it is this force that becomes effective when the birds dive into the water in a nose-dive. Gannets are said to attain—shortly over the water surface—a speed of up to 120 km/h. With angled wings the animals plunge into the water from a 15 to 35-metre height. The fall is accelerated by added wing-strokes. At the same time in order to obtain stabilization of the body the latter is set into rotating motion round its longitudinal axis which is even enhanced close over the water surface as the wings are held tight to the body (thus preserving the rotation impulse). Moreover, this helps to acquire a more favourable aerodynamic shape. Yet, in spite of the streamlined shape, there are considerable forces acting upon the bird's body on its impact against the water. Everybody knows from a "belly clap" how hard the rebound on the water really is. To protect themselves against injuries due to these forces the birds are equipped with a unique shock-absorbing system. A great many gas vesicles distributed inside the skin and between the skin and the muscles act as shock absorbers. They afford the birds effective protection during their ostensibly foolhardy diving manoeuvres.

58 Even the thick-set shape of
the grampus (killer whale), the
largest dolphin, hints at its being
a fast swimmer.

An extreme instance of adaptation to water, the source of nourishment, is to be seen in seals (suborder Pinnipedia). These are beasts of prey whose aquatic way of life is most strikingly expressed in the structure of their limbs. Their story is at the same time one of spatial and ecological isolation in the course of the Tertiary and the Quaternary, which is, however, but incompletely borne out by fossile finds (cf. in E. Thenius and H. Hofer, 1960). The original forms are land beasts of prey which may have gone over to living in water as early as several millions of years ago. There are three groups differing in the degree of their adaption: walrus (family Odobaenidae), eared seals (Otariidae) and common seals (Phocidae); the latter are the most specialized, the former primitive, walrusses being nearer to the eared seals.

Edge-free streamline form allows the animals actually to cut water. Inequalities are filled with cutaneous fatty layers. The neck is mostly drawn back into the trunk, the outer auricles are reshaped, nose and mouth can be shut. Central parts of limbs are shortened so as to form fins, the hind limbs usually being far more strongly altered: the foot has become a bilateral symmetrical paddling surface with webs between the toes.

The primitive eared seals—these, together with the droll sea-lions belonging to their species, are well known to everyone from the zoo or the circus—derive their under-water propulsion predominantly from the up-and-down rowing movements of the finlike enlarged paws with their thumb edge turned downwards. At the same time the feet are stretched backwards and steer the diving seal.

In nearly all other Pinnipediae shoulder and hind limbs play only a slight role, or none at all, as propulsion organs. In them likewise the hind extremities serve primarily as steering devices. It is the dorsal muscles that cause trunk twisting to the sides and drive the sea dog forward.

Deep and endurance diving has also elicited particular adaptations with regard to oxygen storing and the water pressure. The number of red blood corpuscles is generally high, and the blood-vessels tend towards plexus (interlacement) formation. The water pressure is held back by compressible cushions of fat. During the dive the enforced stopping of respiration retards heart-action without causing the blood-pressure to fall. In a sea dog the cardiac rate of 110 beats per minute was found throttled down to 10. The veins of the skeleton muscles and of kidneys narrowed down and yet the brain and the heart continued to receive their normal blood supply. The markedly lower blood supply of the muscles reduces their temperature, which in its turn cuts down their oxygen consumption. All this results in the astonishing diving times and depths in which the supposed record is held by the stately seal, the North Atlantic hooded seal *(Cystophora cristata)*: from the remains of food found in the stomach it was concluded that the diving depth was up to 600 metres. The duration was estimated at about 20 minutes. Since seals generally catch their prey rapidly and consume it on the water surface, the diving times find their limits in the kind of nourishment consume: it is certain that the natural performance capacity is often not exhausted. However, 100 metres depth and 10 minutes diving time are nothing out of the ordinary for many fin-footed animals.

59 Slim torpedo shapes of porpoises (or bottle-nosed dolphins) *(Tursiops truncatus)* during a simultaneous leap.

60 The faces of these fast swimmers are marked by concentrated attention.

61 The slim body profile of this emerging fin whale *(Balaenoptera physalis)* but slightly ruffles the water surface.

62 Typical of diving ducks is the deep position in water and the fact that the legs are fixed far at the end of the body. This is clearly visible in the European pochard *(Aythya ferina)*.

63 So akward on land, so elegantly glides the ringed seal *(Phoca hispida)* through the water. Every "braking" accessory of the body has been "retracted" by evolution.

64   Its spindle shape, favourable
from the streaming-technical point
of view, allows the swimming
penguin as it were to fly through
its medium.

65 *Fulmarus glacialis*, the Holarctic fulmar petrel, is easy to tell from any gull by the stiff way it holds its wings in the sailing position.

Flying

In the course of the historical development of the species in the animate nature flying had been "invented" several times: winged insects populated the forests of the Carboniferous period 280 to 300 million years ago, pterosaurs soared into the air about 200 million years ago, bat-like animals have flitted about in the dusk for 60 million years, and birds, too, have been proficient in flying for about 25 million years.

As to Man, the air space has remained closed to him until the most recent past. Full of envy he had to look on while eagles and vultures were sailing ostensibly without effort in their element, swallows swift as arrows were chasing insects, and stormy petrels withstood hurricanes on the high seas as child's play. He himself continued to be fettered down to the earth's surface—and thus to exclusively two-dimensional motion. It is therefore only too understandable that at least in his legends and fairy tales the heroes were able to fly, whether with the aid of sophisticated technical devices like Daedalus and his son Icarus in Greek mythology, or on the backs of mythical superbirds.

However, with the growing of scientific knowledge and the development of technology there have been—starting ever since medieval times—increasingly numerous attempts by bold and imaginative men to force open the gate to active flight by Man. Yet all the efforts—by Leonardo da Vinci (1452–1519) up to Berblinger, the well-known tailor of Ulm who in 1811, amidst the laughter of spectators, crashed into the Danube with his imitated ornithopter—were doomed to failure, having exhausted themselves in mere outward imitation of the flight of birds. Man's anatomic peculiarities and, above all, the physical laws of air flow past a flying body, remained unobserved. Nor could they have been taken into account, since in most cases they were as yet entirely unknown. Of course, as far as Man's anatomic prerequisites are concerned, there had already been A. Borelli (1608–1679), professor in Florence and Pisa, who had calculated that man's pectoral muscles would have to make 17 per cent of the weight of his body should he wish to rise into the air by means of wing strokes. In actual fact, however, our pectoral muscles are weakly developed and their performance amounts to about 1 per cent of the body's weight.

Though all flyers among animals have a very light structure of the body, buoyant lift, whose decisive influence for the water fauna we have come to know in the preceding chapter, plays no role here in view of the low density of the air (0.0013 g/cm³). However, it has of course its peculiar place in the history of flying, for the first free man's flight was carried out on the "lighter-than-air" principle.

The historic event took place on November 21st, 1783, in France when two daring men with a hot-air balloon of the Montgolfier brothers rose into the air and flew over Paris at a considerable altitude for twenty-five minutes.

66 The legendary flight of Daedalus and his son Icarus. According to the legend Daedalus made wings for his son and himself out of birds' wings and wax in order to be able to escape from a prison in Crete. During the flight Icarus approached the sun out of curiosity. The wax melted, the wings fell down, and he found his death in the sea.
After an old German woodcut from G. Wissmann.

It was Otto Lilienthal (1848–1896), the outstanding pioneer of flying who—together with his brother Gustav—were the first to investigate scientifically the flight of birds and the air flow past a wing. In so doing, he also came to appreciate the importance of the arched profile. His book "Bird Flight As a Basis for the Art of Flight" ("Der Vogelflug als Grundlage der Fliegekunst"), which appeared in 1889, made him well known and famous in the expert world. On a plain girding machine, which had constituted the most important equipment for flow testing as early as in the 18th century, the brothers measured aerodynamic forces acting upon the wing at different speeds and different angles of attack. The aerodynamic lifting force and drag were entered into a polar diagram, a kind of representation which is still usual in aerodynamics even today. This aerodynamic lift is of quite a different character from the static one that we have already got acquainted with; it is linked to the streaming media. In the air at rest—while this means being at rest in relation to the flying object—no aerodynamic lift occurs. Hence the assumption suggested itself that the lift must be associated with pressure conditions in the air flowing past the airfoil profile—but how? — Daniel Bernoulli (1700–1782) had discovered that there are two different kinds of pressure in flowing liquids—static pressure p and dynamic pressure q, while the latter increases with the square of velocity v:

$$q = \frac{\varrho}{2} \cdot v^2.$$

Both are interrelated according to the equation discovered by him and called the Bernoulli equation $p + q = p_o$, p_o being the static pressure in the liquid at rest and thus a constant quantity. Accordingly, if the velocity and, quadratically with it, the dynamic pressure increases, then the static pressure p must decrease. Luckily, the equation is valid not only for liquids but for the air as well as long as the velocity of flow lies deep enough below the velocity of sound (round 330 m/s).

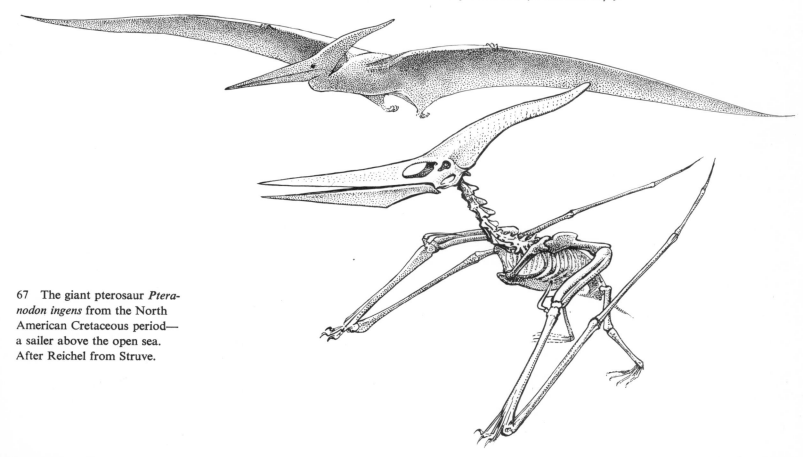

67 The giant pterosaur *Pteranodon ingens* from the North American Cretaceous period—a sailer above the open sea. After Reichel from Struve.

Static pressure p_O is the generally known air pressure of 1 kgf/cm². As for conditions at the profile it follows from the Bernoulli equation that although in the places of a high rate of flow a considerable dynamic pressure is bound to occur, it is accompanied by a low static pressure. A lower static pressure in relation to the environment entails suction effect. Every reader is familiar with this suction effect of rapidly flowing liquids and gases from the water-jet pump or the Bunsen burner, in which the natural gas streaming from the jet sucks in the air for combustion from the environment. The hydrodynamic paradox is also a generally known phenomenon. The air blown from a jet against a plate does not possibly push the latter before it but, on the contrary, the plate is attracted, as the air streaming off on the sides parallel to its surface causes a drop in static pressure.

Let us, however, turn back to the flow at the aerofoil. Pressure conditions—and hence the aerodynamic lift—are particularly easy to explain on a profile with the clear flow pattern: the particles in the streaming air move along certain lines—the stream lines. If during the flow round a body these lines lie close together, then the velocity of flow is high, if on the other hand they spread farther apart in other places, then the rate is lower. In the first case it is an underpressure and in the other case an overpressure that occur in relation to the environment. One can see on the airfoil profile that the stream lines are closely compressed on the convex upper surface while on the concave bottom surface the distance between them is larger. As a result, there is bound to be a light underpressure on the upper surface and a slight overpressure under the profile. Both pressure deviations operate in the same direction and bestow lift upon the aerofoil. At the same time, out of the total lift about two-thirds are accounted for by the underpressure and only one-third by the overpressure. So just like aircraft birds, too, are drawn upwards rather than being lifted by the air flowing under the wing.

Of course, in absolute terms the deviations in pressure from that of the ambient air are slight. Thus, even under the airfoils of an air-liner weighing many tons the overpressure is less than 1 per cent higher than the ambient air pressure. A common buzzard with a surface load (= body weight divided by wing surface) of a mere 0.4 gramme weight per square centimetre manages an increase in pressure by about 0.01 per cent.

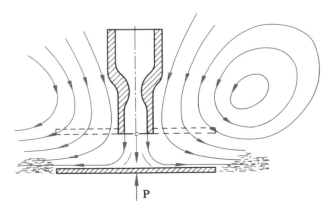

68 The hydrodynamic paradox. The plate placed below the nozzle is not possibly pushed away by the fast escaping air but rather pulled upwards. The reason for this is the lowering of the static air pressure above the plate as a result of the high velocity of the air streaming off along the sides. A second plate drawn in a dotted line would even substantially increase the suction effect.

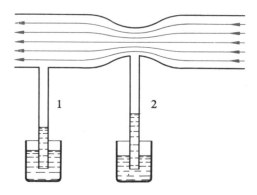

69 Static pressure in streaming air. In streaming fluids and gases the total pressure p_0 is composed of the static pressure p and the dynamic pressure

$$q = \frac{\varrho}{2} \cdot v^2$$

(Bernoulli equation:
$p + q = p_0 = $ const.).
p_0 is a constant quantity, namely the static pressure in the medium at rest. The fluid columns give the static pressure in the upper flow tube. Through the narrow spot the air flows faster than in the rest of the tube, the dynamic pressure being higher there in consequence. The static pressure, on the other hand, gets smaller since the sum of the two pressures is constant. Hence in the tube of pressure gauge 1 the fluid is sucked in more strongly than in the tube of pressure gauge 2.

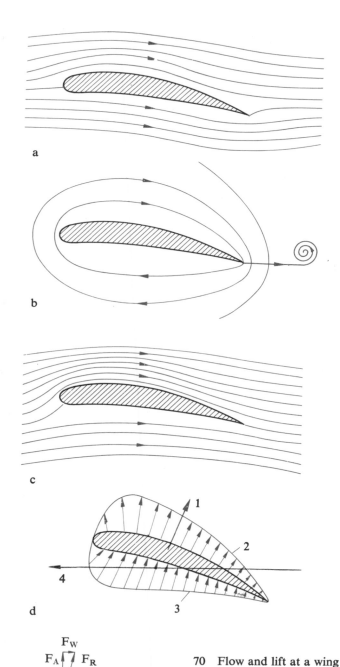

Lifting force is calculated by using the relation

$$F_A = c_A \cdot \frac{\varrho}{2} \cdot v^2 \cdot A,$$

which contains dynamic air pressure

$$\left(q = \frac{\varrho}{2} \cdot v^2 \right)$$

[c_A = lift coefficent, a dimensionless figure dependent on profile, surface condition, and angle of attack, against the streaming air; A = total surface of the wing]. An entirely analogical formula yields the aerodynamic drag:

$$F_W = c_W \cdot \frac{\varrho}{2} \cdot v^2 \cdot A$$

[c_W = drag coefficient] which occurs in the opposite direction to that of the motion. This is a formula we have already come across in the introductory chapter entitled "Swimming". It may equally be used to calculate the descent velocity of a parachute jumper.

Besides lift, it is also propulsion that an active forward flight of a bird or an aircraft requires. While the bird produces both lift and propulsion by using its wings, in an aircraft with its rigid airfoils forward thrust is produced by propeller or jet propulsion. Then lift comes about as a result of the high relative speed between the aircraft and the ambient air at which the flow pattern described above arises on the airfoil profile.

70 Flow and lift at a wing section.

a—potential flow during which no lift occurs.

b—circulatory flow which is created simultaneously with an initial eddy that must, for physical reasons, turn in the opposite direction. The initial eddy "swims" off backward.

c—potential flow plus circulatory flow result in the actual lift-producing flow at the profile.

d—pressure conditions at the profile and the resultant aerodynamic force (1) in the thrust point. The arrows point in the direction of the acting forces.

In sphere 2 there is light underpressure, in sphere 3 light overpressure as against the undisturbed environment. Arrow 4 indicates the movement direction of the profile in the air.

e—forces at the profile: G weight, F_A aerodynamic lift, F_W air resistance, α angle of attack, F_R resulting force from lift and air resistance, 5 thrust point, 6 chord. After B. Eck.

Flying Facilities in the Vegetable Kingdom

In the organic world air movements are a "favourite" means of conveyance, and the number of tiny germs of vegetable and animal life that are borne away by each blast of wind by far exceeds our imagination. In particular it is pollen, seeds and fruits that are propagated by the wind and engage our attention—for who has not, at least on one occasion, stood on an elevation and observed how the warm gusty wind balls up golden clouds out of myriads of microscopic pollen corpuscles above a pine-grove in bloom. A similar thing can be observed above blossoming fields of corn, and how you can powder yourself yellow all over is well known to everyone who has wandered through blooming hazel-nut shrubs. With many plants pollen is transferred by insects, and nature has been exceedingly lavish in providing structures and facilities for this purpose. Anemophilous plants, on the other hand, ensure pollination by exposing the pollen-pregnant anthers to the moving air so that the pollen can be easily shaken out.

Pollen grains are small and are thus eminently suited for transportation by air. Sizes and shapes of course vary and have no small effect on the behaviour of pollen in the air streams. Thus small pollen grains usually drop to the ground more slowly than the larger ones in little moving air; since, however, varying surface structures can be observed there are numerous deviations from this rule. Even within the same species not all pollen grains are equally good or bad "flyers"— their sizes often differ, and at night when the air is particularly still it could already be established that the smaller pollen grains of one species were floating at a greater height than the others. This is to be ascribed to their lower descent velocity. Naturally such kind of delayed dropping favours a farther conveyance, and this can sometimes be an advantage in respects of selectivity for the continuation of the species.

In daytime aerial bodies are mightily intermixed. This, for instance, causes the fact that the air over a forest is almost equally interspersed with pollen up to the height of several hundred metres. "Pollen has been found at the height of 2,000 metres and even more above the ground ... in a certain period 2.7 millions of oak pollen grains driven by the wind flew through every square metre of the air surface daily. Those are noteworthy figures; it is hardly at all possible to form a correct idea of the wealth of pollen in the air on a good pollination day." (cf. in W. Jacobs)

There are also seeds and even fruit bodies whose minuteness alone enables them to be carried away by the wind from their place of origin. Thus we know seeds of orchids of which a million grains make up 1 or 2 grammes. Hence one seed would weigh 1 or 2 microgrammes, and it is easy to imagine that these "micrograin flyers" can under certain conditions fly over considerable distances, are borne up into high altitudes, and that their low descent velocity guarantees wide propagation. The seed-bud and storage tissue remain weakly developed, air-filled cavities inside and a large surface reduce density and multiply quite considerably the travelling capacities of these structures. Thus everything is so designed as to enlarge the body surface as compared with the volume, to offer much effective surface to the air flow, and at the same time to preserve the extreme lightness of the whole thing.

The descent velocity of the light pollen has once again something to do with the laws of flow in gases. As the tiny particles descend very slowly and thereby produce no whirl in the air, it is possible to apply just for calculating this velocity the Stokes law of resistance for viscous (laminar) flow. Already after a brief period of descent the velocity remains constant— air resistance and pollen weight keep balance:

$$F_W = G = 6 \cdot \pi \cdot r \cdot \eta \cdot v_s$$

[r = pollen radius,
η = viscosity constant of the air = $1.7 \cdot 10^{-5}$ kg/m $\cdot$ s,
v_s = velocity of descent].

If the formula is adapted according to velocity after inserting mass times volume times acceleration due to gravity for the weight then the following formula is obtained:

$$v_s = \frac{2 \cdot r^2 \cdot \varrho \cdot g}{9 \cdot \eta}$$

[ϱ = density of the pollen grain,
g = acceleration due to gravity = 9.81 m/s^2].

The descent velocity is all the lower the smaller the radius and the smaller the density.

How minutely beyond human imagination plants are able to dimension their propagation corpuscles can possibly be illustrated by a glance at mushroom spores (cf. H. Hertel, 1963): With a radius of 5 μm and a density of 1.4 g/cm³ they have a weight of merely $5 \cdot 10^{-11}$ gramme weight (thus 20 thousand million spores go into one gramme). According to our formula the spores descend with the minimum velocity of 0.0045 m/s. Even the minutest vertical air motion tends to elevate them and takes care of their spreading over vast areas.

For the most part we come across spherical seeds in "micro-grain flyers" though reticulated reliefs on their surface in the form of elevated ridges already hint at the evolution step which we subsequently find implemented in the shape of the extraordinarily manifold "flying organs" of flying fruits and seeds. "The sphere is a body which possesses the smallest surface with the largest volume. Any deviation from the spherical shape with the space volume remaining the same leads to an increase in the surface and hence to a retardation of descent in the case of a descending body." (cf. in W. Jacobs, 1938)

The flying facilities of plants are generally divided into hair-like and winglike ones. The popular symbol of the former is the universally known "puffing flower", the matured syncarpy of the dandelion *(Taraxacum officinale)*. If one blows strongly into their tender hairy parachutes the individual stemmed tiny fruits get torn free and drift away in "stable equilibrium". The hairs are always hollow and filled with air—hence their lightness.

On the other hand, the winglike facilities are richly represented already in our European flora; many forest trees send their seeds and fruits on the journey equipped with supporting wings, propeller blades, etc. At the same time the focus may be placed quite differently (same as with the "hairy" flyers) depending on the particular place where the "nut" (embryo plus stored provisions) happens to lie. In accordance with this and the wing design (there may be one wing or more in existence) varying descent motions occur. These flying facilities have very little weight. They are made lighter by built-in air-filled spaces even if the former sometimes do reach a certain size.

According to W. Jacobs in the case of the maple fruit the organic substance of the wing has a density of 1.5 g/cm³ but the wing as a whole about 0.46 g/cm³. "That such light and therefore delicate wings do not tear at the edges is safeguarded by an economical yet extremely ingenious distribution of the supporting tissue." As a rule, "autorotations" round the heavy core are carried out and glidings are performed on a more or less narrow spiral as trajectory path. The inert human eye perceives this rapid helical motion about the vertical axis as an area of circle. The air streams through this plane as a hose from below upward, the flyer preserving inherent stability by the rotation and braking the fall to a maximum degree. Besides, in the same way a descending helicopter with switched off (or stalled) motor is slowed down. According to H. Hertel, given a strong wind (16 m/s) and a tree height of 10 metres the fruit of a flyer performing a helical motion and descending at the velocity of 0.8 m/s is borne over a distance of 200 metres.

The rotating wings have a rigid structure and permanent shape. The propagation of seeds is possible even without flowing or moving air: The fruits of the Indonesian climber shrub *Zanonia javanica* perform glidings which had enraptured—shortly after the turn of the century—the Etrichs, father and son, who after the death of Otto Lilienthal both carried on his flying-machine experiments. They are—technically speaking—all-wing gliders. The slightly upturned wing covers about 5 centimetres times 15 centimetres, yet the whole structure weighs a mere half a gramme. The centre of gravity, i.e. the boss, is situated close to the leading edge and actuates a comfortable gliding flight on a spiral path.

Let us return in our minds once again to the "puffing flower". Does not a comparison with parachute jumpers just force itself upon the mind in our highly mechanized world? Under the carrying structure which in descending offers great resistance to the air hangs the actual useful load—here the seed body, there a human being. Of course, for calculating the velocity of descent we must now apply the quadratic law of resistance, for its substantially higher value as against that observed in the case of pollen results in different flow conditions.

In the case of balance, weight and air resistance are once again equal to each other so that for velocity of descent we obtain the following formula:

$$v_s = \sqrt{\frac{2 \cdot G}{c_w \cdot A \cdot \varrho}}.$$

According to this equation, a parachute with a surface area of 50 m² brakes the fall of a man weighing 75 kilograms down to the constant velocity of round 5.3 m/s. If there were no braking air resistance then a pilot who jumps off his aircraft in the altitude of 1,000 metres would, in accordance with the law of falling bodies, reach the ground at the final speed of 140 m/s (about 500 km/h)! However, this speed is not really attained, since the falling body with a surface of about 1 m²

has an appreciable air resistance. If this resistance is taken into account in the calculation, then the impact velocity is "only" about 130 km/h.

The idea of floating safely to the earth by means of a braking umbrella from towers or ledges is no doubt very old. In Peking as early as in 1306 on the occasion of an emperor's accession to the throne Chinese jugglers are reported to have jumped off high towers with parachutes to amuse the populace. In 1495 the versatile Leonardo da Vinci worked out a project of a parachute containing a total of 36 square metres of sailcloth. Fauste Veranzio, a Hungarian mathematician, is said to have jumped off a tower with a rigid parachute of an 8 metres times 8 metres surface. Even before having invented the hot-air balloon the Montgolfier brothers occupied themselves likewise

71 The parachute jump of Fauste Veranzio (Hungarian mathematician) about the year 1617.
From H. Buch and D. Strüber.

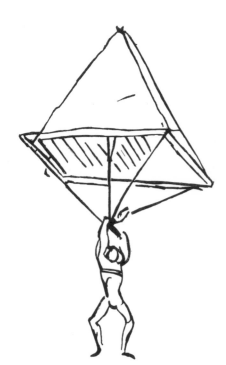

72 Self-portrait of Leonardo da Vinci (1452–1519).

73 Parachute design after Leonardo da Vinci.

74 The way primeval birds flew, has not yet been fully clarified. The nearest explanation is probably a combination of successive flapping the wings and gliding. Here *Archaeopteryx siemensi* in the Berlin Museum of Natural Sciences.

75 Otto Lilienthal during his gliding attempts on the *Maihöhe* at Steglitz in the spring of 1893. Above, he has just jumped off the roof of a flight shed. The bottom picture shows Lilienthal trying to prolong the path of flight by putting his legs sideways at an angle.

76 A portrait of Otto Lilienthal, the German pioneer of flight (1848–1896).

77 In late summer the winged fruits of maple-trees spin down to the ground.

78 Natural flying equipment of *Dipterocarpus spec.*

79 The "puffing flower" of our young days. Each gust of wind takes with it the ripe little fruits of the dandelion.

81 Diagrammatic sketch of a
parachute jumper. The arrows
F_W and G represent the forces—
air resistance and weight—acting
in opposition to each other.

After a brief falling time, in a
steady state, F_W and G are equal
in magnitude.
After P. Antonov.

with parachutes. In 1777 Joseph Montgolfier (1740–1810)
jumped with a self-built device from the gable of his parental
home and landed safely without sustaining any injuries. In the
19th, and even more so in the first half of our own century,
the parachute had developed out of an interesting technical
toy into an urgently needed last life-saving equipment for the
occupants of damaged aircraft before crashing. Thus, for
instance, in 1785 a parachute brought rescue in the utmost
need to the Frenchman Jean Pierre Blanchard (1752–1809)
when his balloon crashed.

Numerous improvements were invented in the next decades.
The physicist Lalande came to realize that the uncontrolled
escape of the compressed air sideways in the parachute canopy
was the cause of the hitherto inescapable heavy swinging
movements. This evil was subsequently removed by making a
hole in the centre of the canopy through which the compressed
air could escape unopposed.

Jumping from aircraft brought new problems for the parachute designers at the beginning of the present century. It
turned out that in order to guarantee a safe jump the parachute
must be worn on "the man's body" and could not be fastened
to the aircraft. In 1911 G. Y. Kotelnikov, a Russian, developed
a usable life parachute for airmen. In 1913 Otto Heinecke
introduced his well-known knapsack parachute to the public.

Particularly in the Second World War it was the military
people who also took possession of the parachute. Not only
soldiers but heavy equipment and implements floated to the
earth on parachutes. (Of course, to put down a caterpillar
tractor about 4,000 m² of surface are necessary.)

In recent years an important role is played by the parachute
in the safe landing of the returning space probes, let us think
only of the "Soyuz" and "Apollo" undertakings. As a brake
parachute it has been in use for aircraft and even for ships for
many years now.

Vertebrates Demonstrate Gliding

If one is on the look-out for the "flight" phenomenon among the vertebrates one is struck by the fact that apart from birds there are still other groups which have developed the ability to move about in the air space—in a passive as well as active way, such as fishes, batrachians, reptiles and even some mammals. Of course, a qualification must immediately be added to the effect that out of all the groups mentioned above it is only bats and fruit bats (Micro- and Megachiroptera) that are actually capable of an active and protracted flight. All the others which we are about to discuss here demonstrate merely the most primitive and simplest form of flying, the gliding, and a more or less perfect one; for the moment we leave out of account the whirr of the silver hatchetfishes (Gasteropelecidae) which is very limited in both space and time.

The "classical" flying fishes of the tropics—some 30 species—all belong to the Exocoetidae family. Travel stories often record encountering them: "Ship passengers can hardly cross the Atlantic Ocean without having representatives of the best gliding species *Exocoetes volitans* fly on their deck even if the ship's hull projects high out of the water. They are lifted up by the upwinds flowing past the vessel's walls. Now these fishes have not acquired the ability to fly perhaps because they derive pleasure from it, but as a result of a selection process, of a selection pressure which had been instigated by the pursuers of the Exocoetidae. These are primarily the dorados which chase them in a mad pursuit. The flying fishes try to save themselves by long gliding flights, rising repeatedly into the air, though in the end exhausted from the rush they must dive into water again where the less dexterous fall prey to their pursuers. Besides the dorado there are also other predatory fishes and birds lying in wait for the flying fishes." (K. Deckert, 1967)

Among these most primitive vertebrates we know "two-winged" and "four-winged" animals; the former use only their long thin-skinned pectoral fins as rigid gliding surfaces, the others (genus *Cypselurus*) spread at the same time their equally oversized ventral fins and in this way carry out a gliding flight on the biplane principle. Essentially the "ascent" of these silver-scaled fishes takes place as follows: under water the fish takes the "run-up". Then its body rises obliquely out of the water with a parallel spreading of the pectoral fins. The caudal fin (whose submerged lower lobe is distinctly longer than the upper one) beats mightily towards left and right until the animal, after about 20 metres of "take-off" acquires the speed of 20 metres per second. At this speed (about 70 km/h) the body shoots at an angle of about 30 degrees against the horizontal line out of the denser medium and describes now a slowly sinking trajectory of a few dozen up to at the most a 100-metre distance which ends in plunging into the water. However, by a number of fresh take-offs (again the mighty, accelerating tail thrust in the water with the trunk gliding

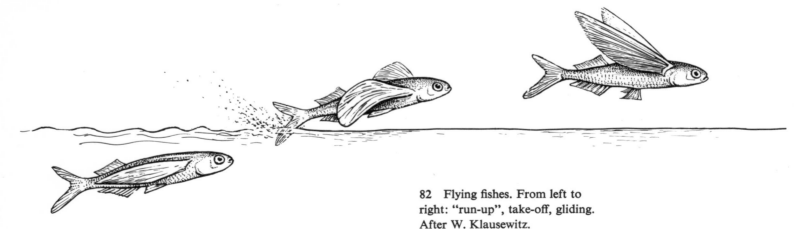

82 Flying fishes. From left to right: "run-up", take-off, gliding. After W. Klausewitz.

83 *Racophorus reinwardtii*,
a Javanese flying frog.
After E. Haupt.

horizontally over the surface) it can be prolonged to 300 or even 400 metres. Normal high flights (steep trajectories) are rare; a maximum of 8-metre flying altitude has been observed.

The recent (now existing) batrachian and reptile world can boast only of relatively little developed flying performances, and it is often difficult to decide whether one has to do with a parachute or a gliding flyer. In this we follow J. A. Oliver (1951) who conceives as a parachute flight a retarded falling whose deviation from the vertical line does not exceed 45 degrees. If the former is larger than that it is a gliding flight that occurs.— For instance, the "flying frogs" (genus *Rhacophorus*) indigenous to the Indonesian island world brake down the fall with unusually large webbings between fingers and toes and skin edges on the forearm and foot. The animals spend nearly all their lives in the tree-tops of their tropical habitat; to this extent their "webbings" have undergone a remarkable change of function. R. Mertens, the renowned Frankfort herpetologist, reported in 1959 on this aspect about the relevant experiments carried out by W. Senfft with the Javanese *Rhacophorus reinwardtii:* "Senfft made his frogs jump from about a 3-metre height, and in so doing the widely spread webbings attracted his attention; however, the landing proceeded too quickly to allow of a more exact observation. Therefore he ascended the roof of his garden hut and now he forced the frog to jump from a height of about 4.5 metres. After about 2 metres one could notice a distinct decrease in the fall due to the extended, in upward direction slightly arched toe webbings. The frog's body seemed to be hanging on four bright-coloured parachutes.

84 Repeated take-off and flight
of flying fishes.
After W. Klausewitz.

Several repetitions established beyond doubt the fact that our frog, though incapable of a gliding flight with a slow descent, was really capable of a fall-diminishing parachute flight. The path of flight described a parabola which, after a 1.5-metre fall bends steeply, almost vertically, to the ground. During the air trip the extremities are constantly bent at the same angle to the body . . .".

In the twenties a landing of the South-American *Hyla venulosa* 31 metres distant from the vertical line was observed by H. B. Cott; the frog having jumped down from a 42-metre height.

If we go one evolution step farther, we encounter a small South-East-Asia agama which demonstrates the most perfect gliding among present-day reptiles: the flying lizard or dragon lizard (genus *Draco*). Its "parachute" are creased skins on the sides which are spread out when the flight begins. The spreading is managed by a number of "false" ribs, i.e. ribs not connected with the breast-bone, which are considerably elongated and are moved by the intercostal muscles.—Let us once again listen to R. Mertens (1959): "As I was watching, on a tree-trunk in the botanical gardens in Bogor on the island of Java, an eagerly dozing male flying dragon *(Draco volans)* stretching out the yellow wattle, there flew from the same spot another flying dragon that I had overlooked: the first impression was one of a large nocturnal butterfly! The little animal, a female, landed with its head raised on a neighbouring tree 10 metres away. Hardly had I had a look at it than there followed also the first discovered male animal in a gliding full of drive, landed close to the female and at once embarked upon a graceful wooing game by bowing and several times repeated displaying the laryngeal appendage. Then the wooing ended and the little animal pecked eagerly at the numerous horse-ants, its chief nourishment."

Flight distances of 20 metres and even a little more have been noted. As a rule, the flight trajectory deviates only 20 to 30 degrees from the horizontal line. A small climb rating caused by upwind has also been observed.

By the way, it might be mentioned that the spread flying membranes in these graceful reptiles as movement organs are there not only to increase air resistance but also serve as a menacing means of defence against enemies and as a wooing "stunt" in mating.

Finally, such leaping gliders are also to be found among recent mammals: marsupial mammals, rodents and the "big gliding flyers" of the order Dermoptera. Together with the vegetarian cobegos or flying lemurs (Cynocephalidae), of the Malay Archipelago, this comprises only one family with two species of the genus *Cynocephalus.* The animals are something quite specific within their class, and until 30 years ago, since the days of Linné, they had been frequently "shoved about" a lot by the perplexed comparative anatomists and taxonomists (cf. in H. Petzsch, 1966). These animals, too, prolong their jumps to a significant degree thanks to a finely haired flying membrane which almost entirely enfolds the body from the neck to the tail. If the animal is resting in the tree-tops in its characteristic suspension on all fours with the back outwards or downwards, it is enveloped by this patagium as though by a coat cut far too wide (only finger and toe claws remain free). Gliding paths of about 70 metres have been known. The animals, once they have dared to jump from high in the air, generally land on the lower stem of a neighbouring feeding tree and quickly proceed to climb up to its fruit-laden top. Finally, there are "flying" rodents which we find among the Pteromyinae, the flying squirrels, and which are to be found all over the world with the exception of the Neotropics and the Arctic zones. Their furry flying membranes are comparatively less developed but receive support from a cartilaginous or osseous sickle-like appendix on the wrists (flying squirrels) or on elbows (scaly-tailed "squirrels"). In the case of flying squirrels gliding paths of 30 metres have been measured.

The gentle reader who has the fancy for such things is hereby invited to follow us in his mind on a warm summer-day on a ramble through a landscape in which the hill alternates in a variegated sequel with expanses of fields, meadows and pasture land and little copses; thus through a landscape as we know it from many a place in Central Europe. A sonorous "Hiya" suddenly comes from somewhere, and yet we are unable to discover its originator. Finally, he appears in our field of view, there, over the edge of the forest a medium-sized brown bird of prey is rowing comfortably in our direction whose flying image is well familiar: *Buteo buteo*, the common buzzard, a bird which as an eager hunter for small rodents plays an important role in the household of nature surrounding us. "A few rowing movements, then a gliding forward, now a turn and a wonderful hovering in spiral lines without beating the wings with widely stretched flight feathers . . . a piece of magic, of poetry, perhaps even lovelier than the song of the lark and the singing of the nightingale!" With so much feeling did Otto Kleinschmidt, the outstanding German ornithologist, describe the little spectacle of nature, enacted at this moment before our eyes, already a good many years ago. And it is indeed ever anew a fascinating sight to watch how such a miniature eagle pulls himself up without so much as a wing-beat into heights in which he then merges with the sky as a tiny spot still visible with the naked human eye.

Our buzzard may serve here as an example for eagles, vultures and storks in other climates. These often represent the scene just sketched above in a still more impressive manner— no wonder when it is borne in mind that for instance the largest bird of prey on the Earth, the condor *(Vultur gryphus)*, this powerful new-world vulture, has a wing-spread of 3 metres. One is apt to wonder spontaneously what natural equipment enables these birds to perform such feats, and, furthermore, what air forces and movements carrying the bird are here at play.

First, it is to be remembered that the bird's body is structurally built so as to make use of air space. The capacity for a free movement in the air medium with its low density (0.0013 g/cm³) enforced specific adjustments, particularly of the skeleton and in the muscular system in the course of the development of the phylum. A system such as the skeleton which naturally and in an easily comprehensible way tends towards being heavy was "lightened off". Of course, not by simply reducing the osseous substance during the phylogenesis but by pneumatizing most of the bones, as well as by the fact that those spaces which, in mammals, are filled with heavy stuff, such as marrow, are hollow and air-filled in birds. Other bones again show a sponge-like design. A number of air sacs, appendages of the lungs, extend as far as into the larger bones. Despite all this, such an extraordinary strength and loading capacity is demanded by this supporting system which must constitute a reliable abutment for the lever arms of the wings moved by considerable forces. This is guaranteed by a strong admixture of built-in salts in the bone substance and by special connections between the respective partial systems.

At all events and taken all around, the bird's body as compared to its size is light. Nevertheless, the buoyant lift, which in water with its relatively high density (1 g/cm³) is of decisive importance for all living creatures when swimming and diving, plays no part in flying. A bird becomes a flyer and a sailer only thanks to the surfaces that keep it up, i.e. the wings. Their skeleton axes consist of the same parts as the human arm, only the "hand" has undergone a strong involution. These axes stabilize, reinforce the forward edge of the wing. In the backward direction stretches the actual wing surface formed by tightly packed longitudinally arched arm wings (secondaries) and rather more transversally arched primaries. Finally the compactness of the surface is made complete by short and long deck feathers (quill-coverts), and these are responsible for the typical wing section, convex on the top side and concave on the lower surface. They serve to smooth the wing surfaces the way one can see on birds in general that all irregularities are smoothed out by the feathering, all accessories and appendices being drawn into the feather cover. The streamlined body and the wings offer the air only a small resistance, thus being flowed past in a laminar, and hence eddy-free, way. Flow and pressure conditions at the profile and the aerodynamic forces associated with them have already been described in the introductory chapter "Flying" where the formulas for their calculation have also been given.

The lifting force, the resultant of many components of forces at the profile depending on the flow pattern, acts on the centre of pressure and operates vertically to the direction of flow in an upward direction. The vector addition (parallelogram of forces) of drag and lift yields the resulting aerodynamic force that is directed diagonally backwards and upwards. What acts as a third force in the centre of gravity is weight. This will pull the wing downwards in the direction of the centre of the earth. Therefore only at a sufficiently high speed the bird (or the aircraft) receives enough lift not to drop down. Without propulsion, which is just what this forward speed is produced by and what overcomes air resistance, only a gliding flight is pos-

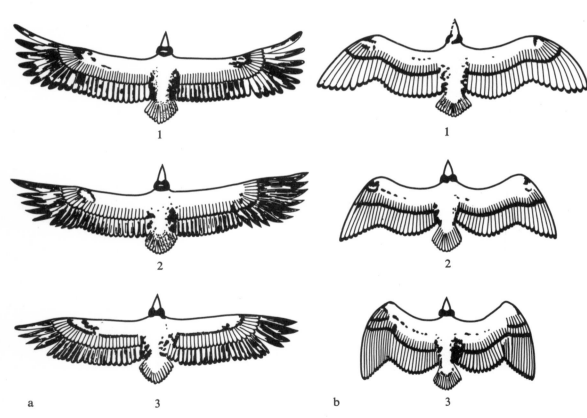

85 Variable wing top view in various flight conditions of the vulture.
a—vulture in sailing flight:
1 Rising,
2 Flying horizontally,
3 Descending.
b—gliding flight:
1 Flat,
2 Normal,
3 Steep.
After H. Hertel.

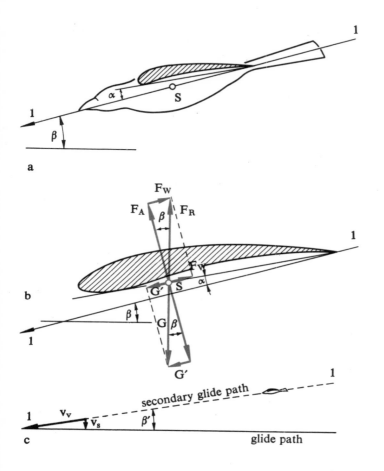

86 Gliding flight in a stationary state.
a—bird in a gliding flight.
1,1 gliding direction, α angle of attack of wings, β angle of glide.
b—forces at the profile in stationary gliding. The resulting aerodynamic force F_R and the weight G cancel each other out, likewise the components of both of these forces. In stationary flight forward speed can be calculated by equating F_W and G':

$$F_W = \frac{\varrho}{2} \cdot v_v{}^2 \cdot c_W \cdot F_A = G'$$
$$= G \cdot \sin \beta.$$

c—glide path and secondary glide path.
The descent velocity v_s equals forward speed v_v times $\sin \beta$.
After K. Herzog.

sible in an oblique direction downwards. The path that can be covered in this case by gliding vertebrates or sailplanes from a certain height down to the landing point can be calculated from the glide coefficient, which equals the quotient from drag and lift, i.e.

$$\frac{c_W}{c_A}$$

(dynamic pressure and area cancel each other out). Very good gliders among birds attain glide coefficients from 1:17 (eagles) to 1:20 (albatrosses), while good sailplanes 1:40 to 1:51. The glide coefficient means that for instance during a calm, an eagle can cover a straight-line distance of 17 kilometres by gliding, i.e. with no wing-beat, from a height of 1,000 metres.

Then how is it possible that the bird or sailplane in spite of the braking air resistance hover through the air spaces circling gracefully for minutes or even hours, that without a single wing-beat or any propeller drive they not only do not sink down but rise to considerable heights and cover many kilometres?

It is the upwards directed currents of air, the product of certain landscape structures and weather conditions, that starting with the softest whiff of a wind (v lower than 1 m/s) are skilfully utilized by the bird as well as the flyer of the sailplane as sources of energy for the climb. On this point let us listen to the captivating story of the "Bearded Vulture in the Himalayas" by Bengt Berg: "Under the rocks, about 1,000 feet high, where I was lying in hiding there stretched for miles a valley punctuated by ravines between narrow mountain ridges and slowly sinking into the misty blue depth, far away where the Sutley river ran under the snow-capped mountains in the west ... Above a gorge down below I saw the copper colours of the first bearded vulture (lammergeyer) glimmer against the sun during a turn. In wide circles he was gliding slowly forward. He was evidently making a search of his hunting preserve. He glided into a ravine and out again, turned round the rock edge into the next crevasse, searched it and then did the same with one after the other, always in the same way in spirals, circling upward ever higher above the valley. My eyes were so enthralled by his motions that I forgot to look at the watch to see how long the circling lasted. More than a quarter of an hour must have passed when all of a sudden he, more rapidly than ever before, approached the steep mountain-side where I was waiting. In this time interval he seemed to be making not a single wing-beat, and still he was soaring upward all the time. As he reached my steep mountain-side, the rock ravines stopped, and there was obviously nothing for him to search for.

In a straight line along the rock he came soaring up to me, just at such a distance that the mountain ridge below me did not hide him from my sight as he . . . rushed past me. He went into a climb which would have been steep for a horse on the ground, and so he flew past and vanished by a uniform flight far overhead where the steep mountain-side turned a corner. And until he disappeared—still without any wing-beat . . .

Outside over the valley, far and near, dark-winged Bengal vultures and mighty isabelline Himalayan vultures were circling above . . . They came sailing along one after the other as if pulled by some invisible strings through the valley close to the mountain crest, the same way the bearded vulture had come before them and disappeared high above in the mountains. Not a single one came from a different direction . . .

It was a magnificent spectacle, and it began to dawn on me that this was a veritable vulture route, where the only thing one needed was to lie down in wait with the camera and to operate it correctly at the right moment to obtain all the pictures one might wish for . . .

How was this possible? And how did it come about that coming from the same direction persistently he took that same certain course. As I rose to climb on and felt once again the pleasant draught of air streaming up from the abyss down below I said to myself that there could be no other more valid reason than this: The bearded vulture knew well where the heated air throughout the day was streaming upwards along the mountains. He, and with him the other . . . flyers had their course where the current of air was bearing them. There he had just as much certain paths in the aerial spaces as men and goats had their own here down below on the ground slopes, though these did alter with the winds and the times of day."

Berg's "Himalaya Dragon" sailed in the upmoving wind, the best known kind of upward oriented airstream. The rock slope forced the horizontally streaming air mass to turn aside vertically, and the gliding vulture, by means of his broad wings at any rate little loaded per unit area, managed to convert the "upward component" of the air compensating or even surpassing his descending velocity as accessory energy into climbing power, that is to say into flying altitude. Karl Herzog, the indefatigable observer and analyst of biological phenomena concerning flight, formulated this fittingly as follows: "By a minute alteration in the angle of attack or by a weaker or

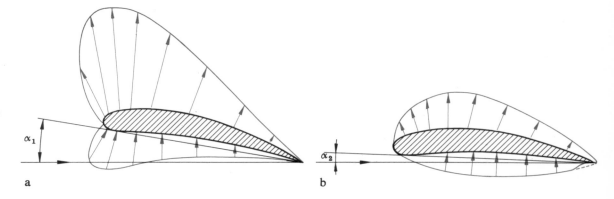

87 Lift and angle of attack. The length of the arrows is a measure for the magnitude of the lifting forces.
a—large angle of attack and hence a large lift.
b—small angle of attack and hence a small lift.

stronger spread of their wings, these masters in the art of flying have the knack of adjusting to the most varied airstreams and exploiting the finest upwinds. As long as it is in any way possible they tack about in such upwind regions, while the wings ... largely take on the task to encompass as much as possible of these ascending air masses ... In so doing the tail surface is twisted to the left and to the right so that the wings remain in horizontal level and obtain the greatest possible amount of ascending power from the vertical current."

The second possibility of soaring is provided by thermal upwinds or warm air currents. The sun heats the earth's surface with differing intensity in view of its changing geological constitution and cover of vegetation. The more heated air comes off the ground surface as a bubble or a tube, pierces the colder air by which it is surrounded and ascends at a speed of several metres per second, not infrequently up to the height of 2 to 3 kilometres. Buzzards, eagles, vultures, storks, pelicans, or sailplanes, too, are borne upward by such thermal bubbles like elevators. When the birds leave these zones, then a gliding flight or a rowing one (active bird flight with 0.5 to 3 wingbeats per second) takes them into the next ones.

The pilot of a sailplane, the technical opposite number of soaring birds, has no energy source in his aircraft for propulsion at his disposal. He gets himself pulled up into the height by a cable winch or by a power tow plane, and once the tow-cable has been released, he has to look around quickly for suitable upwinds if he does not wish to have to come down back to the earth immediately by a gliding flight.

Since the days when in 1911 the history of (German) gliding sport began with gliding flights performed by Darmstadt pupils and students on the slopes of the Rhön (mountain range in Bavaria, Federal Republic of Germany), amazing flying performances have been accomplished by the great aces among sailplane pilots. In 1951 Carl Erik Oevergard, a Swede, pushed on into a height of 16,800 metres on the Long Wave, an upper air current over the Sierra Nevada. His aircraft having no pressurized cabin, he suffered the high-altitude death. As early as 1941, while exploring the lee wave over the Gross Glockner (highest point in Austria) the German Erich Klöckner had reached 11,500 metres. Nor have flights extending over many hundred kilometres been a rarity any more for a long time now. A distance of 749 kilometres was covered by the Soviet woman sailplane pilot Olga Klepikova in 1939. On July 31st, 1964, the American Alvin H. Parker surpassed the 1,000 kilometre mark by 36 kilometres in a flight lasting 10.5 hours.

88 Two different kinds of ascending currents for the soaring flight.
Left: ascending heated air masses (thermal bubble), right: hill-side upcurrent. With a hill-side upcurrent horizontal winds are directed upwards along buildings, hills or mountains and thus receive a vertical component.
After K. Herzog.

In order to exploit and master the variety of different up-winds one must possess excellent meteorological knowledge and gain a lot of experience. The fact that even the rapid up-wind before a thundery front is suitable for high-altitude flights was unwillingly discovered by the German Max Kegel, thereupon called "Gewittermaxe" (i.e. the Thunderstorm Max). Yet woe betide him who finds himself in the turbulences with their upward and downward currents inside a thunderstorm!

There are dangers menacing the pilot even in the finest weather. As already mentioned, lift and drag depend on the angle of attack between the section chord and the direction of the oncoming flow. If initially both of these increase with the angle of attack, above about 15 degrees the lift decreases and

the drag is greatly increased. This is what happens in the flow pattern: the flow hitherto laminar turns turbulent. Advancing from behind to the front eddies form—the flow tears off, and this causes the lift-producing flow pattern to disappear. Then the aircraft flies in a "stalled" condition and pancakes or "nose-dives". This situation is one that every bird avoids by instinct.

There is nothing more impressive beyond reality—this is what was experienced by Maja Schiele, an enthusiastic sail-plane-pilot, during a gliding flight in the French Alps: "Ever more overwhelming is the view through the cabin bonnet not iced—thank God! From the North East to the North West ranks one mountain range on to another range: innumerable white, jagged tops drive up skywards out of the blue-grey

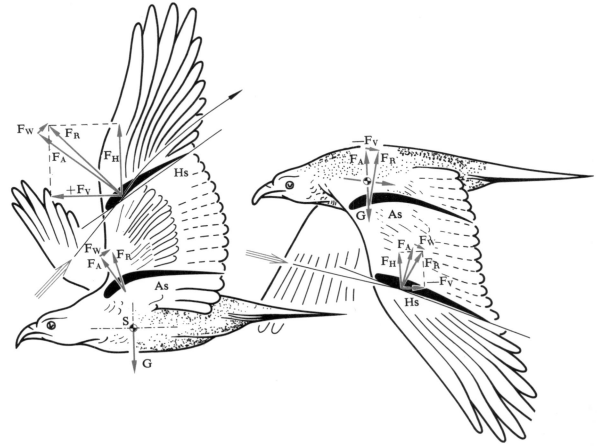

89 Propulsion and lift during bird flight. The left picture represents the downstroke. In the various zones of the wing both propulsion F_V and lift F_H are produced. F_A lift of the profile, F_W resistance, F_R resulting aerodynamic force, As secondaries' section, Hs primaries' section. After K. Herzog.

91

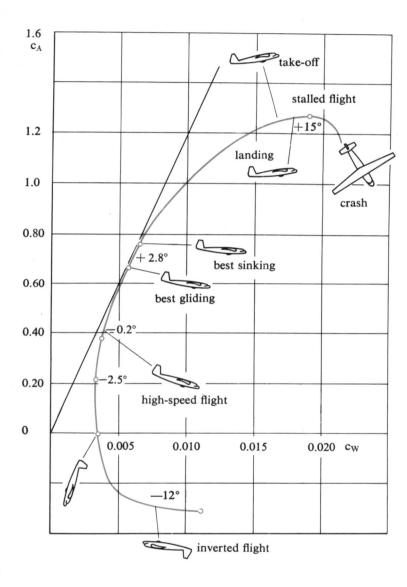

91

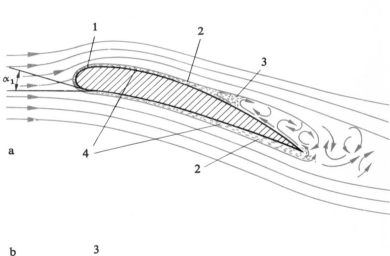

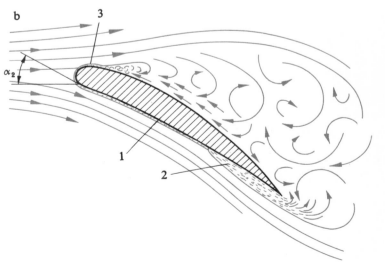

mountain massifs. In the South there is the sea, standing out distinctly from the grey mist, the Bay of Marseille, some 150 kilometres away. Remarkable to be so high up, at 6,000 metres surrounded merely by a little plywood! At once magnificent and anguishing. Ever anew the same tactics: a narrow full circle and push scanningly to the left or to the right against the wind. Maximum climbing 1.5 m/s. Now only a few needle breadths to 7,000! ... Must not become careless. With a heavy heart I say *Adieu!*, good-bye to these world-distant altitudes, to this unreal space in which one could stand amazed and dream away for hours."

90 Polar diagram after Otto Lilienthal for a sailplane. The data of the angles on the curve give the angle of attack of the profile. The entire curve is obtained in such a way that the pair of values c_A (lift coefficient) and c_W (drag coefficient) are established for all possible angles of attack. After M. Schiele.

91 Detachment of boundary layer at the profile at too large angles of attack α.
a—beginning of the detachment.
b—complete detachment in a stall.
1 Laminar boundary layer,
2 Turbulent boundary layer,
3 Detachment point,
4 Transition point.

Dynamic Sailers above the Oceans—
Stormy Petrels and Albatrosses

Now that we have got thoroughly acquainted with the static flight of the large birds of prey, storks etc. above dry land let us still turn to the sailers of the high seas. Above the world oceans, the "weather kitchen" of our planet, there are latitudes both on the Northern and the Southern hemisphere with extreme wind conditions in which these graceful sailers are at home. The wind conditions—wind is streaming air which moves in accordance with laws of physics from spheres of a higher pressure to those with a lower one—are entirely different above the sea from those prevailing above dry land. Wind and the water surface are in close mutual interaction. Above the waves, themselves a result of air movement, there arise not only upward and downward currents on their windward or lee side respectively which are, moreover, accompanied by a perceptible change in the horizontal velocity of the air currents, the velocity decreasing in the direction towards the water surface. In these complicated circumstances with wind speeds amounting to several dozen m/s a flight referred to as dynamic sailing is performed by albatrosses, stormy petrels and frigate birds. They utilize the rapidly altering wind speeds over the waves to glide over the oceans for hours, yes, even days without any bird flight—for which many of them are not particularly well suited or equipped. In so doing they mostly sail close above the water surface and do not allow themselves to be borne up higher than 20 or 30 metres. This is to be understood, since in the higher altitudes the local alternation in wind direction and wind speed—the prerequisite for a power-economizing sailing flight—is absent.

Their nourishment, which consists of cuttlefish, crabs, fishes and ship's waste must be picked up by the high-sea flyers from the water surface. Being ideal sailers and lightly constructed, they receive a strong buoyancy in water, and are therefore very poor divers.

When flying vultures and albatrosses are compared with one another, the difference in the flight pattern is obvious to everybody. Vultures have planklike broad wings, while those of albatrosses and petrels, on the other hand, are narrow and tapered. This is the adjustment to the different wind conditions above land and above the sea. Let us recall for a moment: land sailers are marked by low wing loads, sail slowly, and thus require none too high a rigidity of their wings. The air flows past them with relative calm and uniformly, at any rate, on a tempestuous wind they are hardly to be seen describing their circles.

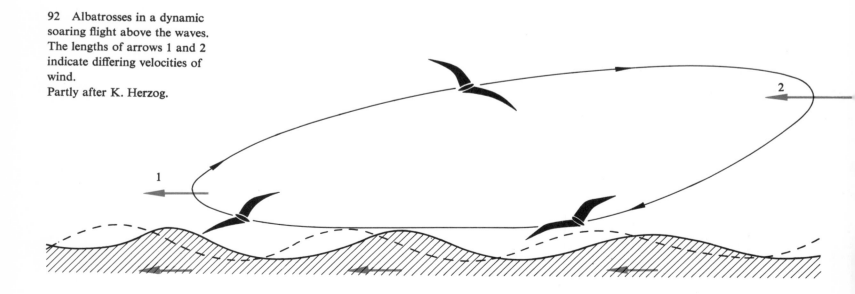

92 Albatrosses in a dynamic soaring flight above the waves. The lengths of arrows 1 and 2 indicate differing velocities of wind.
Partly after K. Herzog.

High-sea sailers who spend the major part of their lives in flight are exposed to a by far higher loading acting on their narrow wings. Their profile with the small drag endowing them with strong lift allows them to reach significant gliding speeds, providing them at the same time with a considerable torsional strength of wings.

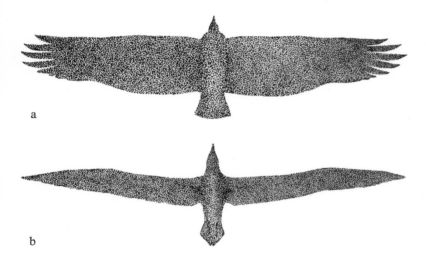

a

b

93　Wing contours of a land sailer (a) and a sea sailer (b).

For a better illustration let us now quote the account by the Frankfort zoologist W. Klausewitz with *Fulmarus glacialis*, the fulmar petrel: "A few hours after Snaefjellness, the white landmark of the Icelandic West coast, had disappeared from the horizon, the weather changed all of a sudden ... There rose a strong wind with the corresponding waves. The ship is now surrounded by swarms of birds. They are fulmar petrels which not only follow the steamer at a small distance but are sailing round it so close as to give rise to fears they might get caught and remain hanging in the rigging. Their wings are constantly stretched, and while showing no bird flight, they stay in the air for hours. They seem to be spending no additional energy during the flight.

If they drop down on the water ..., they subsequently have no manner of difficulties with taking off again. To rise up in the air again, the birds simply allow themselves so to say to be thrown up into the air by the next wave crest: the moment they reach the wave top, which may be several metres in height, they spread out the wings against the wind and let themselves be lifted several yards into the air by the upwind characteristic of each higher wave. After a light circle downwards the bird strikes the upwind of the next wave and is catapulted a considerable bit of a distance higher.

The whole of this manoeuvre is carried out without any kind of wing-beats in nothing but a gliding flight.

Every wave with a distance of 50 to 100 metres between the crests and a height of between 6 and 10 metres has its own little wind system with a downwards gliding stream on the lee side and a powerful upwind on the windward side ... Having allowed itself to be carried upwards by the upwinds, up to the culminating point and having suffered a considerable loss of speed, the fulmar petrel makes a turn and sails very steeply down, thus attaining a high flying speed ... Subsequently it frequently glides over and along a wave crest while making constant use of the side current as upwind. After a certain distance ... it once again turns against the wind, thereby obtaining the necessary lift to raise it to the culminating point of its trajectory. In this way the fulmar petrel flies for rather a long period and covers rather a long distance without ever having once moved its wings. This flying technique, and the physical constitution associated with it, in particular the shape and the functional performance of the wings, are so higly specialized that a full utilization is possible only in a stronger air motion. The flight of this bird is less adjusted to conditions when the weather is calm and the sea surface unruffled."

It is obvious that the largest sea birds, the albatrosses of the predominantly southern latitudes, likewise need rather a fresh wind to perform their fascinating flying manoeuvres. If the wind drops they no longer manage to follow the ship they are travelling with by means of sailing, and thus often fall behind. The largest albatross species is the wandering albatross *(Diomedea exulans)* with a wing-spread of between 3 and 4 metres —something unique in the recent feathered world.

In principle, these feathered giants use the air streaming above the water in a similar way as their small cousin from the Northern hemisphere, with one possible exception, i.e. that the trajectories of the latter are not so distinctly marked as those of the former. Though these characteristic paths (cf. Fig. 92) make for a significant prolongation of their flight, they are nevertheless described from "purely aerodynamic reasons in the sense of the power-saving sailflight" (W. Klausewitz, 1971).

The wind systems above the waves and further up, where their speeds rapidly increase, are used over and over again. This flight is variable to the extent as it is, logically enough, modified by the changing weather conditions, though what happens is in principle ever the same: the albatross sails parallel to the ship's course, the wind blowing at a right angle to it. The bird heaves to and in consequence of gravitation glides under the wind ever more downwards with increasing speed. Close above the water it makes a 90-degree turn and now glides with the cross wind between two wave crests. Whereupon it turns once again against the wind, and the upwind of the wave crest drives it more and more upwards whereby its forward speed decreases and the kinetic energy acquired during the downward glide gets "spent up". At the climax of this trajectory our long-winged wanderer heaves to again and now follows the vessel once more for a bit. Then if it becomes too slow it must once again gather momentum, and thus the downward gliding is repeated. So far man has been incapable of accomplishing the dynamic flight of the high-sea sailers with his sailplane.

The Hummingbird– a Miniature Helicopter

Hummingbirds, those minute flying jewels, are dispersed over the New World in about 320 species. Once they had made accessible to themselves as a source of nourishment the wealth of blossoms of tropical regions as far as high up into the mountains and had gradually become adjusted to it functionally. The name refers to the humming noise made during the flight that is not infrequently given off by these tiny birds.—"Here stands a tender creature as if motionless in the air before a blossom, its long thin bill dipped in the nectar cup; only a grey flitting veil betrays the fact that the tiny wings are actually engaged in furious activity. Then the hummingbird, constantly holding its body in an almost vertical position, dissociates itself from the blossom—and this is done backwards! A spirited bend, then it "stands" again for a few pulses motionless in the air: syrphids (or hover flies) can do this in a similar way. Up and down hovers the hummingbird, backwards and forwards— something that hardly another bird is capable of, nor can it be imitated by any aircraft in equal perfection. However, our admiration for the hummingbirds' art of flying is tinged with a feeling of having seen a thing of magic beauty." We owe this beautiful description of the graceful flying manoeuvres of those feathered dwarfs to the Berlin ornithologist G. Mauersberger.

In short, what we are concerned with in this whirr is an extreme vibration flight. The wings also swing in a horizontal line only forwards and backwards, but they are at the same time so twisted in themselves that during the forward strokes one can see from above their upper sides, and in the backward strokes it is the lower parts that are displayed to the view (cf. Fig. 103). "The action is to be equated to an alternation of a forward and backward flight, the irregular twist (aerodynamic twist) of the wing corresponds to a propeller whose driving force pulls in a vertical direction and accords lift just as while flying in a horizontal line it affords propulsion. Seen from the side, the wing tip describes a horizontal eight." (J. Steinbacher, 1959)

In an age of technology what suggests itself directly even after these few lines on the flight of the hummingbird is a comparison with a helicopter. Even though the flying manoeuvres of the latter are hardly so lovely and graceful there are no great differences in principle: vertical take-off and landing, hovering flight, flying forward and backward, these accomplishments have been mastered to an equal degree by the helicopter just as much as the hummingbird. However, the similarity is not limited to manoeuvrability. Both flying machines, the living and the technical one, function in an analogous way: they produce lifting jets which give them the necessary lift. In a helicopter this is taken care of by one (occasionally even two) horizontally arranged airscrew, the rotor. The hummingbird reaches the same effect by swinging its wings backwards and forwards. Free rotation about an axis, one of the most important forms of motion in all fields of transport and mechanical engineering is impossible in animate nature—owing to insuperable difficulties, e.g. in blood supply and nerve connections.

94 Comparison of wings between a hummingbird (a) and a bird sailing overland, the buzzard (b). The part of the primaries in the entire wing is substantially more strongly marked in the hummingbird than in the buzzard. After H. Hertel.

Rotor blades as well as the hummingbird's wings have the airfoil profile already familiar to us so that the lift developed during rotation or wing-beat is to be explained in the same way as that on the airfoil during a forward flight. Once again there develops on the convex upper side light underpressure and overpressure on the lower side. These pressure conditions also make it easily understandable how a lifting jet occurs, i.e. the air is sucked in from above by the underpressure and accelerated in a downward direction by the overpressure. By means of the properties of the lifting jet it is possible to calculate power

expenditure during the hovering flight when the lift equals the weight of the flying object. The air streaming downwards through the rotor or the hummingbird's wings has an impulse $m \cdot v$ [m = mass of the streaming air, v = its velocity]. The temporal alteration of this impulse equals the sought lifting force for which, after few transformations, the following relation is found:

$$\varrho \cdot v^2 \cdot A_s = F_A = G$$

[ϱ = air density, A_s = jet sectional area].

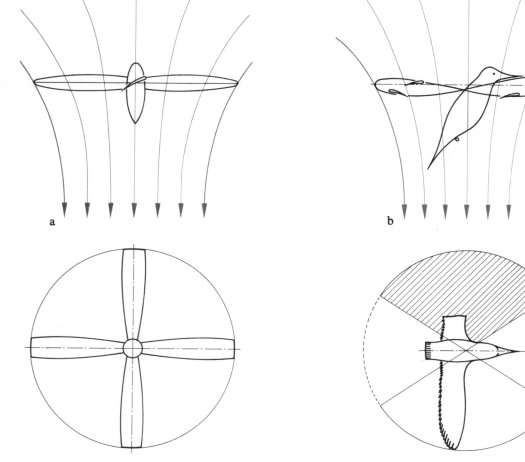

95 Jet lift of a helicopter (a) and of the hummingbird (b). The propeller rotating in a horizontal plane produces an air stream directed downwards, which according to the reaction principle evokes the upward lift of the helicopter. A similar lifting jet is also produced by the hummingbird which of course cannot let its wings rotate round a vertical axis, and therefore it moves them back and forth with an enormous speed in the shaded segment. After H. Hertel.

97 A flying squirrel which covers gliding distances of about 20 metres with the aid of its extended flying membrane on either side of the body.

98 Even the slender form of the osprey *(Pandion haliaetus)* conspicuously typifies the static soaring flight of the large birds of prey above the land.

99 Frigate birds above their brooding place. The slender elegance of their shapes in motion makes us conjecture that like the sailers (family Apodidae) they, too, demonstrate the perfect bird flight.

100 A power-saving dynamic soaring flight is performed by albatrosses and petrels. Here *Diomedea irrorata*, the Galapagos albatross, is shown over the Atlantic.

102   The "Slingsby T51 Dart"
sailplane gliding in a hill-side upcurrent.

101

103 Charm and gracefulness distinguish the flying manoeuvres of the nectar-sucking humming-birds. They represent living helicopters.

104 The "technical" humming-bird functioning in an analogical way—a helicopter in front of the Canadian Mt. Assiniboine.

105 Tender transparency and highly articulated "ribs" turn the wing of the banded dragonfly into a living technical miracle.

By using this formula H. Hertel (1963) calculated for the hummingbird's wing surface loading of 2.5 kgf/m² a jet speed of 2.8 m/s for the region where the overpressure is completely removed. This, of course, is felt by people like us but as a light breath of air.—Moreover, the jet speeds in wasps and bees lie in the same order of magnitude, while helicopters frequently attain far more than 10 m/s.

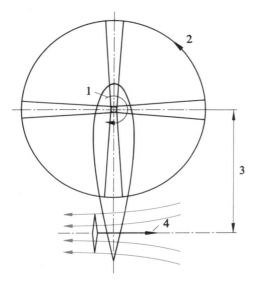

106 Torque compensation in the helicopter. After the propeller has been set into action the left-rotating propeller (arrow 2) makes the helicopter turn to the right (arrow 1). By means of a vertically set little propeller at the rear according to the jet principle a force is produced working in the right-hand direction (arrow 4) which—multiplied by the lever arm 3—produces a torque that prevents the undesirable rotation of the helicopter.
After W. Just.

When wishing to compare the output capacity of various animals or technical structures between them, it is best to go by specific performance (performance divided by body weight). However, in mechanics performance equals strength times path divided by time and hence simply $G \cdot v$. The specific performances then become

$$\frac{P}{G} = v.$$

If one additionally assumes with Hertel that the efficiency of a natural jet apparatus equals that of an airscrew (0.8), then a specific performance of 0.022 hp/kgf is obtained which the hummingbird is able to sustain for several minutes. As related to muscle weight, which makes about a quarter of the weight of the body, the specific performance amounts to 0.088 hp/kgf. This value is about four times higher than that of trained human muscles!—While the high-speed swimmer dolphin, as shown on pages 53/54, has the capacity of maintaining its power input distinctly below the originally calculated high value by keeping the boundary layer laminar, the hummingbird is unable to apply any "trick" of that kind. The hovering flight is physically clear and easy to calculate. Hence the specific performance of the hummingsbird's muscular system is actually considerably higher than that of other birds or mammals. And it is also urgently needed, for the hovering flight demands a far higher expenditure of energy than bird flight or even sailing in which lift is an obligatory consequence of the forward speed. The hovering flight can be performed only by insects with their slight jet surface loading and by the hummingbird.

It is clear to see that such physical top performances have been made possible only by considerable anatomic and physiological transformations. Secondaries were strongly reduced in numbers, and primaries became predominating. The smallest hummingbird species achieve a stroke frequency of 70 to 80 per second—no wonder that the interested spectator is only just able to perceive the wings in action as "grey flitting veils".

106

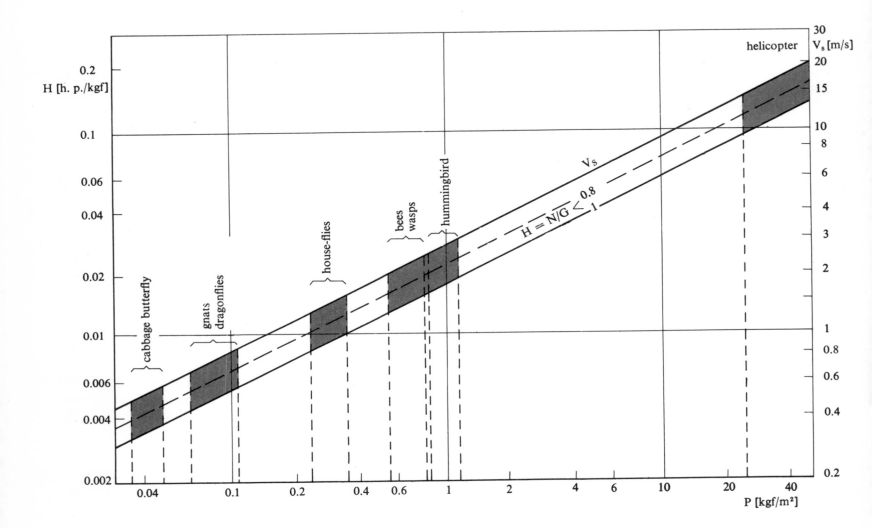

107 Specific lifting capacity H (left vertical axis) and jet velocity v_S (right vertical axis) plotted over the radiant area loading P (radiant area loading = weight divided by radiant area) for insects, hummingbirds, and helicopters. The hummingbird—with an assumed effectiveness of its natural lifting jet apparatus amounting to 0.8—is capable of producing the enormous specific lifting capacity of 0.022 h.p./kgf for several minutes, while the radiant area loading amounts to about 1 kgf/m². It might be that in the case of the "hummingbird-values" the upper limit for the lifting jet flight has been reached in the biological sphere.
After H. Hertel.

Metabolism has increased to a significant degree; in this respect "hummingbird values" cannot be displayed by any other vertebrate. This is conditioned by the nearly constant consumption of large amounts of nourishment, whose transformation into muscular labour demands great amounts of oxygen. This in turn is obtained by faster and at the same time deeper breathing. The oxygen-loaded blood is pumped by a very big (1.8 to 2.8 per cent of body weight as compared with 0.5 to 0.7 per cent in man), and uncommonly fast beating heart into the comparatively "weightier" and specifically more effective flight muscles.

In addition to the whirr on the spot (see above), they have also mastered the high-speed flight. Thus *Archilochus colubris*, the rubythroat, manages 22 m/s, which corresponds to 80 km/h. And the species does need it: being autumnal birds of passage many individuals cross the Gulf of Mexico to Yucatán covering hundreds of kilometres.

For centuries the helicopter principle has played a part in the contemplations of the pioneers of flying. Once again we encounter Leonardo da Vinci who drew sketches of helicopters as far back as around 1500. However, it was only our own century that saw the development of suitable technical conceptions. That is not to be wondered at, as formerly there was no driving machinery of high specific performance at man's disposal. In the years 1910 to 1912 the Russian Yuryev devised the now generally common helicopter principle with a large horizontally mounted rotor and a small vertically placed after propeller serves torque compensation. However, the majority of aeronautical engineers were preoccupied with the airfoil plane in the first decades of the present century, for which they calculated only one-third of the performance necessary for a helicopter of the same size.

In 1937 the first scientifically worked out and fully manoeuvrable helicopter "F61" was introduced by the German Professor Focke, which was once to break all existing records. The specialist world pricked its ears and was partly even sceptical. Two contra-rotating rotors were placed side by side so that there was no need for any additional torque compensation. The speeds reached by the "F61" in straightforward flight exceeded 100 km/h.—For helicopters high forward speeds are problematic for the reason that the rotor-blade tips very soon attain the velocity of sound. In 1963 350 km/h was flown by a French turbine helicopter.

With a helicopter forward flight is achieved by the fact that the rotor level is tipped forward. The jet is then directed backwards and below thus producing both lift and thrust. The angle of attack of the rotor blades and hence the lift can be altered during the flight.

In view of their extraordinary manoeuvrability helicopters have won a wide field of application in both civil and military aviation. They are even used, among others, for erecting transmission towers in a heavy terrain, in mounting aerials on television towers, for rescuing shipwrecked people and for first aid in traffic accidents. Cargo-carrying helicopters transport bulky constructions weighing several thousand kilogramme forces and land them precisely on a narrow space. Equally indispensable are helicopters nowadays in geological reconnaissance or in Arctic and Antarcfic explorations.

What Do Flying Insects Perform?

"When observing insects in flight one finds out that their movements vary a great deal. The cabbage butterfly flutters leisurely over the field, the dragon-fly chases in a zig-zag flight over the pond, remains in one spot, and then darts off forward, sidewards or even backward. The housefly flies round us with a light hum, and the syrphus or hover fly stands with a buzzing flight in one place only to double and fly off all of a sudden. Most varied flight patterns which can none the less be reduced to two types of flight, the bird flight and the whizz which is in fact a most rapid bird flight. A gliding flight gets inserted at the most for a short time and occasionally the hill-side upcurrent can be made use of."

This lively description of flying insects by E. Franz indicates that the animal group which already as to species is hardly to be encompassed has developed hardly a less great variety of flight patterns and motions. Flying speeds, too, show consider-able variations. While the ephemera (or day-flies) reach only 1.8 km/h and hence get overtaken even by passengers, the deer fly reaches 40 km/h, a speed which even trained cyclists find it difficult to achieve. Similar conditions prevail in wing-beat figures: butterflies make as many as about 10 beats per second, flies or gnats, on the other hand, hum about us with a wing-beat frequency of several hundred hertz (1 Hz = 1 circle per second). Many species have a considerable non-stop flight capacity, and thoroughly bear comparison with the flying performances of many birds. Thus migratory locusts often fly hundreds of kilometres and for 20 hours, and the famous monarch *(Danaus plexippus)* migrates in autumn from Canada as far down as to Mexico and perhaps even farther than that, only to migrate back again towards the North the next spring. Flying insects have still been found in the height of 5,000 metres.

This raises the question about the anatomical prerequisites of these impressive conditions. Insect wings are lateral skin protuberances of the 2nd and 3rd breast section. "The skin

108   Flying speeds of a number of insects.

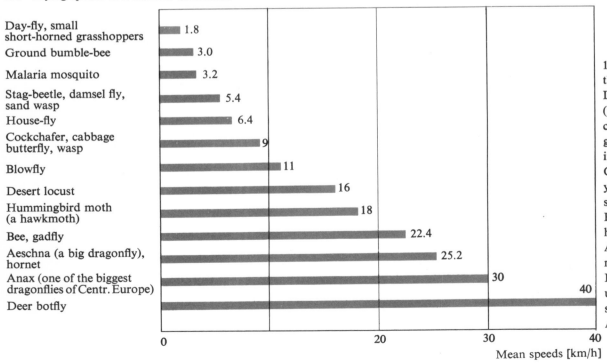

Insect	Speed
Day-fly, small short-horned grasshoppers	1.8
Ground bumble-bee	3.0
Malaria mosquito	3.2
Stag-beetle, damsel fly, sand wasp	5.4
House-fly	6.4
Cockchafer, cabbage butterfly, wasp	9
Blowfly	11
Desert locust	16
Hummingbird moth (a hawkmoth)	18
Bee, gadfly	22.4
Aeschna (a big dragonfly), hornet	25.2
Anax (one of the biggest dragonflies of Centr. Europe)	30
Deer botfly	40

Mean speeds [km/h]

109   Different forms of wings in the insect world.
Left row
(from top to bottom):
caddis fly, shield louse, snake fly, greenfly, soothsayer, twisted-wing insect from the genus *Eoxenos.*
Central row: thrips, fairy fly, yellow jacket, proctotrupid, syrphid, scorpion fly.
Right row: short-horned grasshopper, dragonfly (suborder Anisoptera), damsel fly, hawkmoth, egger, plume moth.
For reasons of book design a unification of the different insect's sizes was necessary.
After W. Nachtigall.

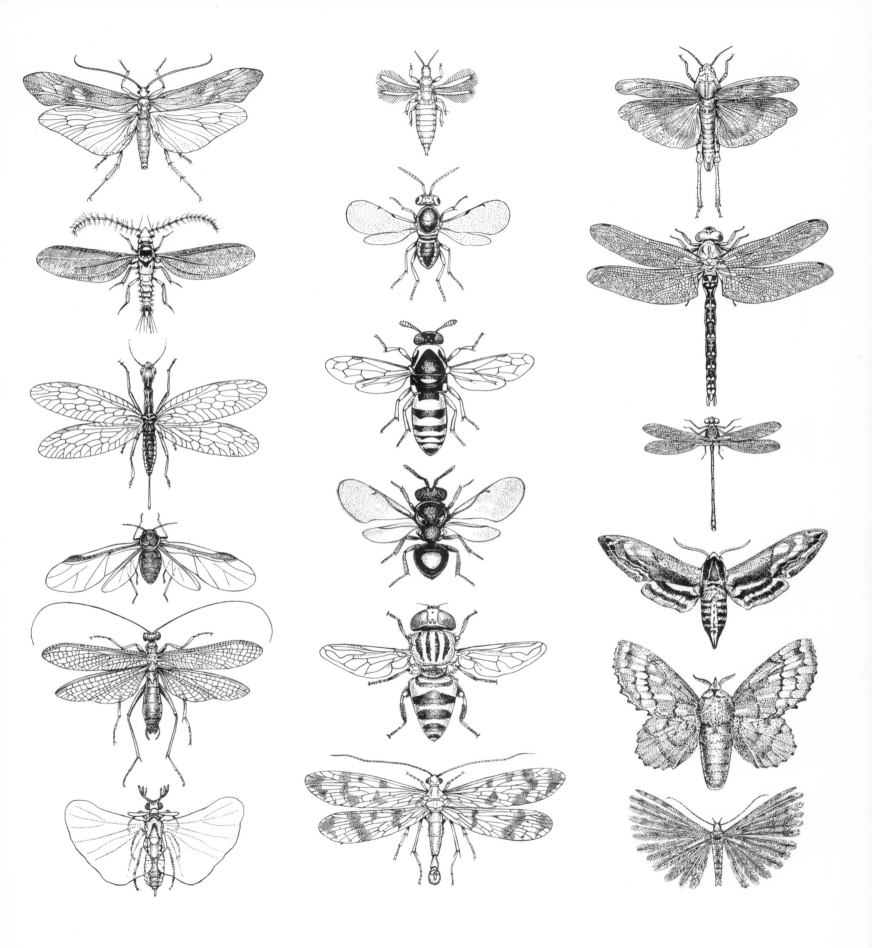

sacs evaginated out of the wall of the body get flattened, the upper and the lower side coalesce in places; from the lower or the higher sheet or even from both of them, 'veins' are formed into which the blood liquid pours and which are provided with windpipes (tracheas) and nerves." (cf. in E. Franz, 1959)

The wing is formed by a double layer of very thin (about 10 μm) chitin (a cell secretion). These surfaces, being exposed to a considerable amount of mechanical stress during the flight, require an effective longitudinal and transversal stiffening, a wing surface framework which can be seen with mere naked eye on resting dragon-flies, hover flies and houseflies etc. as a rich veined structure. This is covered with a wonderfully clear or even colourful wing membrane.

The realization of the wing-beat is a unique phenomenon and is in principle different from that in vertebrates, thus for instance in birds. Since the propping chitin covering—the concurrently protecting external skeleton of the insect's body—lies outside, the powerful flight muscles are fixed on to it from inside. In principle for reasons of mass inertia, it holds that the large insects such as locusts, cockcroaches etc. move their wings more slowly than the small ones. With them the flight muscles act directly on the wing. The central nervous system is in a position to control for each individual wing-beat separately all muscles with impulses. On the other hand, stroke frequencies of several hundred hertz raise considerable problems. In the first place, the· wing mass and area must be small since otherwise considerable acceleration forces would have to be applied. Furthermore, the central nervous system (CNS) is overcharged. It is no longer in a position to communicate separate orders for each individual muscle contraction.

110 Evolution of an insect wing. Section through wing heredity (above) and wing (below). In the embryonic stage two epidermic layers are structured with a ground membrane each (1), these blend (2) into a mid-membrane and then strong skin layers are formed (3). Blood pumped into the intercellular spaces (10) stretches the folded wing.
Epidermic layers and mid-membrane are discarded (3–6). What usually remains are two superimposed skin layers (7, 8).
9 Ground membrane, 11 Mid-membrane.
After W. Nachtigall.

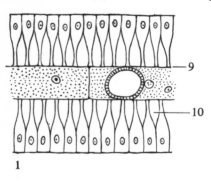

1

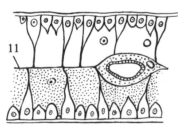

2

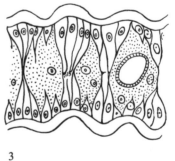

3

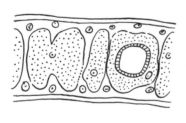

4

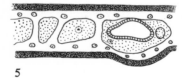

5

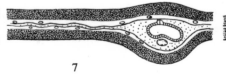

6

7

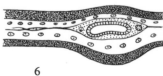

8

What is to be done? Here, too, nature has "contrived" a way out. In flies and other insects with high wing-beat frequencies the wings are no longer driven directly, but indirectly by the flight-muscle system. This happens thanks to the curving and flattening of the thorax to which the wings are attached in such a flexible way that they passively swing up and down between two engaging positions ("clicking mechanism") when the thorax is deformed by the flight-muscle system (see Fig. 111). "The muscles running in the longitudinal direction of the body and those acting transversally tend to curve the chest thus bringing the wing down; others running from the back to the abdominal side contract the chest and thus the wings are lifted." (cf. in E. Franz, 1959) This kind of propulsion then no longer demands that the nerve impulses should necessarily be generated and transmitted synchronically. Rather is the whole mechanism impinged by individual impulses from the central nervous system which occur only at each tenth up to twentieth wing-beat. In the meantime the mechanism swings automatically. In so doing the muscles give themselves the order to contract in the right moment, this happening every time they are jerkily extended by a movement of the thorax. "This is so remarkable and deviates so far from what a 'normal' muscle does that it was not believed for a long time. Since the muscle jerks only when it is mechanically extended, so that the rhythmical movement arises in the muscle (in Greek = *Myon*) itself, this is referred to . . . as a myogeneous rhythm." (W. Nachtigall, 1968) It is this way that allows the impulse frequency limit based on physiological reasons to be eluded.

Flight muscles must be particularly effective. This requires for them to be sufficiently supplied with "fuel" and with oxygen. Only fat and carbohydrates come into consideration as

111 Direct and indirect flight muscles of insects (schematic).
a—direct flight muscles of large insects (grasshoppers, dragonflies) These flight muscles act directly on the wings near the joints.
1 direct wing elevator muscles (contracted in the upper drawing), 2 direct wing dropping muscles (contracted in the lower drawing), b—indirect flight muscles of flies. The indirect flight muscles act on the thorax and not directly on the wings. By the indirect wing elevator muscles 1 (contracted in the upper drawing) the thorax is flattened—the wings are impelled upwards.
The wing dropping muscles pass longitudinally through the thorax and arch it up in the contraction phase (lower drawing) whereby the wings are lowered.
Indirect flight muscles allow of substantially higher wing-beat frequencies than the direct ones. After W. Nachtigall.

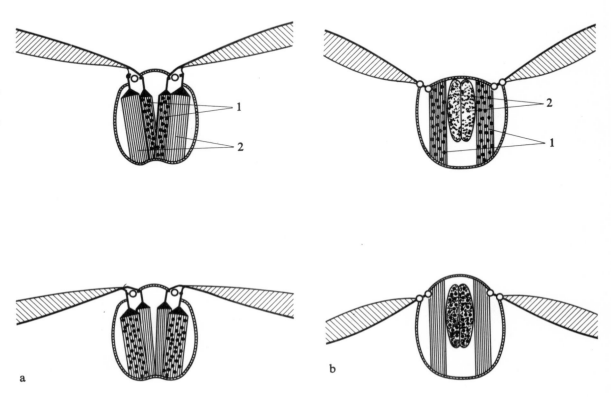

112

fuel, fat being the far more effective of the two. If the flight muscles burn the carbohydrate sugar (glycogen) instead of fat, then the insect when wishing to give comparable flying performances must lug around a multiple weight of fuel. Hence it is no wonder that migratory locusts use the rational fat drive while bees which are able to "re-fuel" frequently are provided with a "sugar engine".

The energy for the muscle labour is taken from the fuel by reacting chemically with oxygen—it is burnt analogically to the propellant in a combustion engine, though of course, at lower temperature. The necessary oxygen is taken from the ambient air, respiration air is brought straight into the cells by a widely extended system of tracheas. Conversely, the combustion product, i.e. CO_2, diffuses from the cell into the trachea. What operates as a pump for aerating the trachea system is mostly the movement of the thorax during flight-muscle contraction. The entire system of oxygen supply is extraordinarily efficient and productive. In 1968 Nachtigall wrote: "While in man a permanent rise in performance can be multiplied at the most five and a half times, the fly raises its basic metabolic rate by the factor 14 ... The factor reached by the bee is 19, the desert locust reaches that of 25, the cockchafer even as much as 107."

Let us add a few words about the flow at the insect wing. Lift and propulsion are produced by wing-beats much the same as with birds yet there are some specific features to be noted which have to do with the small dimensions of the insects. In the introductory chapter "Swimming" we had printed out the Reynolds number

$$\mathrm{Re} = \frac{l \cdot v \cdot \varrho}{\eta}$$

discovered by O. Reynolds in 1883, as the quantity serving to characterize flow. This number can, however, be equally conceived as the quotient derived from acceleration work

$$\left(= \frac{m \cdot v^2}{2} = l^3 \cdot \varrho \cdot v^2 \right)$$

and frictional work ($\eta \cdot l^2 \cdot v$). In the region of large Reynolds numbers it is the acceleration work (inertial forces) that predominates. On the other hand, small Reynolds numbers indicate that more frictional work has to be performed—the viscosity of the medium (the air) acquiring more influence here. Since in small insects both the wing length l and the flying velocity are small, the Reynolds number also becomes small. This means no more than the fact that the tiny flyers already begin to feel the viscosity of the air to a considerable extent. Under this aspect even the bizarre wing forms (such as bristle wings) become comprehensible: the prevailing flow conditions are simply quite different from those obtaining in the case of their big flying cousins or even of birds. In these dimensions profile and curvature of the wings lose their sense, for there arises hardly any lift at the wing. Insects row in the air just as water beetles in the water.

Once again, the example demonstrated on insects goes to show clearly how careful one must be in assessing and comparing properties and performances in various dimensions. The cases when something can be transferred in a linear way are extremely rare.

a

b

113 In the butterfly kingdom multiplicity of forms and colours goes hand in hand with that of flight patterns and movements.
a—Large tortoise-shell,
b—Common tiger moth,
c—Peacock butterfly.

c

114   The small snout pits serving thermal location of prey have given the name to a whole group of vipers. Here this organ is clearly visible in *Trimeresurus purpureomaculatus* between the nostril and the eyes.

Orientation
in the Animal Kingdom

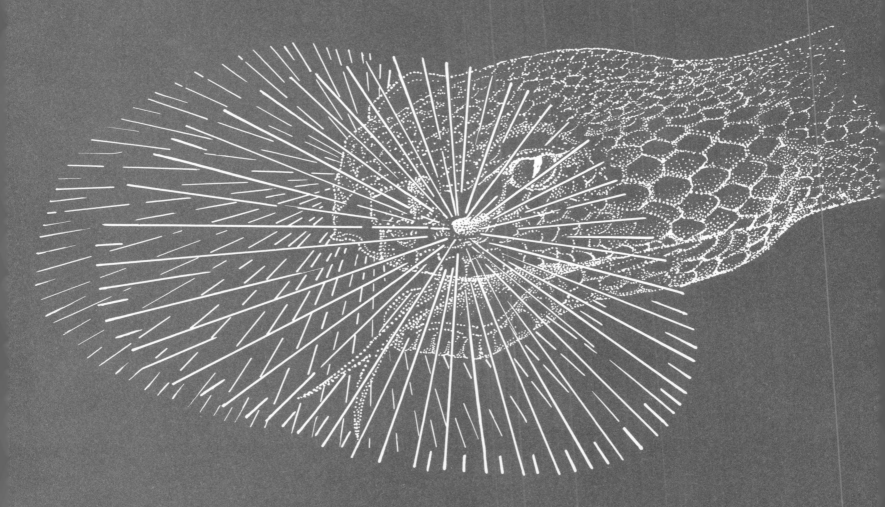

The calm circling of an eagle in the deep-blue summer sky is an aesthetic sight for every man, a symbol of strength, elegance and body control. For the eagle it is simply a search on the look-out for prey. And to do this he must be capable of finding his bearings in space. For him this self-orientation means to find out in what position his own body finds itself in comparison with the reference objects on the earth below him as well as with the horizontal and vertical air currents, and, moreover, in what direction the movement of his body must be made in order that the desired goal—in this case an animal of prey—may actually be attained. The information necessary for the purpose is communicated by the sense-organs (receptors) by means of which orientation stimuli from the environment are firstly recognized and secondly located, over afferent (conducting inward) nerve fibres inside to the central nervous system (CNS) consisting of the brain and spinal cord. There they get processed, and the required orders are passed on to the executive organs (effectors), the muscles, over efferent (carrying outward) nerves. Those sense-organs which are of decisive importance in our hunting eagle are the organ of equilibrium (the labyrinth) in the inner ear, which controls direction finding in the Earth's gravitional field and the changes in the motion of the bird's own body, and the sharp eyes with which he spies out his prey.

With this illustrative and relatively simple example of the problem of orientation one has already raised a number of questions and introduced concepts to be answered and explained respectively. What are actually the orientation stimuli? In what way or ways are the items of information they receive coded in the sense-organs and passed on to the CNS, etc.?

Orientation stimuli can bear an extraordinarily manifold physical and chemical character. There is, however, one thing they do have in common: they convey information with a tiny amount of energy to the sense-organs. Thus the retina of the human eye reacts to a single photon, and the threshold of response of the ear lies for a 1,000 Hz-sound near the unimaginably minute value of 10^{-16} watt/cm^2!—There are sense-organs that react to the gravitational force, such as the already mentioned labyrinth, others register light and analyze it as to direction, colour, intensity or even polarization (e.g. the eye of the honeybee), others again make use of heat radiation, sound waves, within a wide frequency range, the magnetic field, electrostatic fields, the chemical composition of gaseous substances, etc. for obtaining direction-finding information.

Irrespective of the kind of the orientation stimulus all pieces of information—whether in a frog or a man—are uniformly converted into electric impulses of about 0.1 volt and less than a millisecond in duration. Accordingly, it is quite immaterial whether olfactory, taste, optic or auditory perceptions are being passed on, everything occurs by means of the same electric impulses. Only the nerve cords are separated from one another, and flow likewise separately into the brain. Were it possible to exchange the auditory nerve for the optic one, then one would "hear

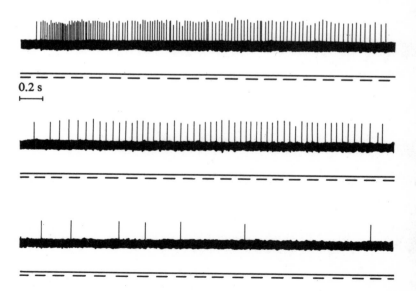

0.2 s

115 The density of nervous impulses of a pressure-reacting sense cell of human skin at varying compressive forces. Above: Compressive force 13 gramme weights, centre: 4 gramme weights, below: 0.6 gramme weight. After Hensel and Boman from B. Hassenstein.

lightning and see thunder" (E. du Bois-Reymond). Thus the receptor converts the continual stimulus into periodic electric discharges, the measure for the excitation intensity being the number of impulses per second—the impulse density. It is to be noted, however, that sense-organs possess the capacity for adaptation. Thus the impulse density passed on to the sensitive nerve fibres depends not only on the intensity of the stimulus but also on its duration. At a constant impulse intensity it decreases with time. For instance, this effect appears conspicuously during the eye's light-dark adaptation. It is for the same reason that we do not perceive continually the pressure of our clothes; after a certain time, the pressure receptors, in spite of the stimulus intensity remaining the same, convey but few impulses per second to the CNS.

Now it is just in this conversion of the most varied stimuli into electrical impulses that the coding of the information in the sense-organs consists. And the code applied is a guarded code par excellence (B. Hassenstein, 1967), for while being passed on over the nerve paths the information does not get adulterated as long as the transmission is not completely interrupted. This is provided

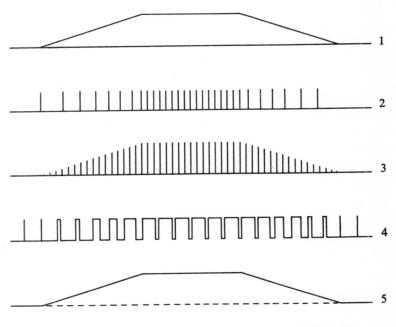

116 Adaptation of a receptor (sense cell) during constant stimulation force. Above: The stimulus sets in suddenly at about 0.75 seconds and ends at 5.2 seconds. Centre: The action potentials (nervous impulses) at first appear in quick succession but then—with the stimulation intensity remaining constant—their density drops to a substantially lower value which also remains constant after a certain time. This process is referred to as adaption and is familiar to everyone from the human eye. Below: Here the state of affairs of the central picture is reproduced in a diagram. The impulse density (impulse frequency f) was plotted with reference to time. After E. Schubert.

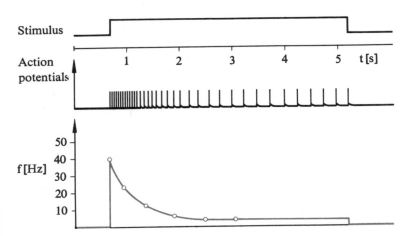

117 Four possibilities of the coding of a stimulus in the sense-organs.
1—the stimulus. To begin with, its intensity increases with time, then it remains constant for a moment, and finally drops again down to zero.
2—stimulus intensity has been coded into electric impulses of equal height and width but of different time spacing.
At 3 stimulus intensity was coded into electric impulses of the same width, of equal time spacing but of different height. Likewise impulses of the same height but of differing width (4) can contain stimulus information. Finally, it would also be conceivable that in the nervous system a direct voltage proportional to the stimulus intensity is passed on (5). Calculations have shown that coding 2, which is the only one encountered in the entire animal kingdom, is least susceptible to incidence of faults and prevents adulterations of information.
After B. Hassenstein.

for by the special irritation and transfer mechanism in the nerve tissue. At each 1-millimetre distance on the nerve paths there are "amplifiers", the nodes of Ranvier, on which each impulse is re-excited. Since the information is only contained in the impulse density and not in the height or duration of a single impulse, nothing can get lost on the way to the brain or to the spinal cord. Therefore the CNS receives exactly the information that the sense-organ has sent on its way.

Orders to muscles or glands are equally given over the nerve paths in the form of electric impulses. (It should not remain unmentioned that there exists still another information transmission—that of the hormones). While in the various receptors physical or chemical stimuli are converted into electric impulses, the command pulses in the muscle fibre stimulate a chemical process which causes their contraction. Here, too, there arise muscle action potentials—likewise electric impulses—that report the activity of the muscles to the CNS. Even in "a state of rest" therefore without outside stimulus operation there is a constant number of impulses per second running over the efferent nerve paths and evokes the muscle tonus. The entire nervous system is characterized by a continuous dense flow of information. Of course, this standing operational readiness of the nerve tissue has been "redeemed" by the loss of the regeneration capacity by division of its cells. Instead of divisibility, the mature nerve cell has only a limited capability of repairing itself when damaged.

Since electricity is the fastest information bearer it hardly occasions surprise that it is utilized also within living organisms for information transmission. However, while technology disposes of metal wires—the best electric conductors—in which the signals are spread by electrons with the velocity of light (300,000 km/s), a living being must "have recourse" to the ion conduction in electrolytes, which in this case represent 0.6 to 4 per cent solutions of ion compounds. Even so this serves to attain a propagation speed of impulses up to 100 m/s, while transmission velocity lies the higher, the thicker the nerve fibre is.

The mechanism of excitation and diffusion of electric impulses in the nerve fibres is founded upon the diffusion of ions. On the nodes of Ranvier the permeability of the fibre membrane for sodium and potassium ions is different and variable. In the non-excited state the concentration of the positive ions—predominant-

ly those of Na^+—is higher outside than inside where the negative ones (e.g. Cl^-) prevail. Hence there exists a voltage of about minus 0.06 V between the interior and the exterior of the fibres, referred to as resting potential. At the excitation of an impulse the membrane suddenly becomes permeable for sodium ions, which then rapidly diffuse into the fibres thereby causing a reversal in the potential, for now it is the positive charges that predominate inside. If the resulting potential of about $+ 0.04$ V (in-

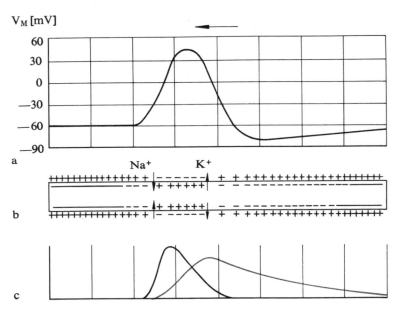

118 Transmission of a nervous impulse along a nerve fibre.
a—nervous impulse = potential difference between the interior and the exterior of the fibre membrane.
b—passing of potassium and sodium ions through the fibre membrane during transmission of a nervous impulse.
c—transmissivity of the membrane for Na^+-ions (left-hand curve) and K^+-ions (right-hand curve) at different places of the membrane during impulse transmission.
After L. Rensing.

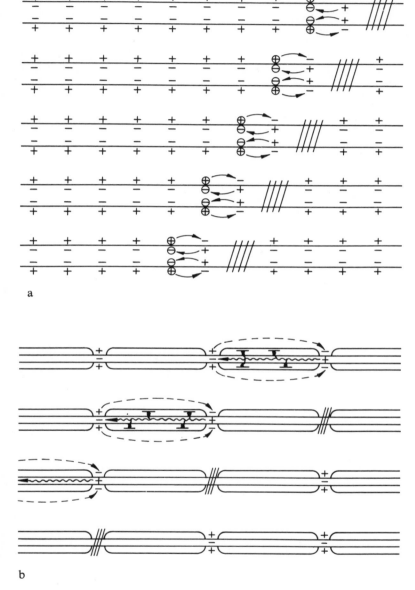

a

b

119 The two forms of impulse
transmission in nerve fibres.
a—nerve without myelin sheathes,
b—nerve with (isolating) myelin
sheathes and nodes of Ranvier.
From E. Schubert.

side as against outside) is related to the —0.06 V in a state of rest the total magnitude received is +0.1 V, which is the magnitude of the nerve impulse. The departing state is reached thanks to the fact that for a time the membrane becomes pervious preferably to potassium ions which again makes the exterior charges predominant. Meanwhile the permeability for sodium has strongly declined. The whole process takes place within less than a millisecond. Between the nodes the fibre is provided with insulating myelin sheathes so that the stream of ions is transferred on over the fibre plasma and over the extracellular liquid. An excitation of the membrane in this area is thus impossible.

In addition to the capability to pick up orientation stimuli from the environment, to process them in the nervous system and to react to it purposefully, some animal species such as the bees and various birds have developed an entirely extraordinary ability: they are able to tell the time very precisely. And no external stimuli are used for doing so! These animals have something like an interior clock. This time-determining mechanism has so far been little explored—thus we do not know for example whether it is a periodic or a continual process that plays a role in this inside the animal's body—the significance of the interior clock especially for the orientation behaviour of the social bees is well known and investigated. Therefore we will revert to this phenomenon in a special chapter.

Nocturnal Ultrasonic Orientation and Echo Direction-Finding of Bats

It is indeed an extraordinary experience of nature to see bats while the night is falling twist round buildings or tree-tops with elegant, at times lightning-speed, turns. Then many an observer may wonder how it comes about that these remarkable mammals who have made the conquest of the airspace find their way in the twilight with such certainty without knocking against something, or even sustaining a mortal injury by bumping against things. M. Eisentraut, the specialist intimately acquainted with the lives of bats and flying foxes (or fruit bats), wrote fittingly about two decades ago that such a question becomes even more pressing when one enters a cave dwelling deep inside the mountains which is never penetrated by a ray of light, in which therefore permanent darkness prevails. Bats will find here the slightest rock protuberance onto which they hook themselves with their feet, or a narrow crevice into which they force their body.

As a result of ingenious experiments, Lazzaro Spallanzani, the outstanding Italian animal physiologist of the 18th century, had proved that it could not be the eyes with which the bats find their bearings during their nightly insect hunting. In 1794 this was a sensational though equally quite enigmatic discovery. The researcher made the animals fly through obstacles in the form of vertically stretched strings and found that they were capable of doing this without touching them. And this astonishing capability was retained even by some which the experimenting scientist had deprived of their sight.

Soon afterwards the Swiss naturalist Jurine undertook a verification of the test results gained by his Italian colleague and obtained an entirely surprising result. After blocking their hearing among other things by plugging their ear canals he found the chiroptera to be all of a sudden strikingly constrained in their sense of orientation.

Unfortunately, these guiding experiments of early empirical zoology remained unnoticed. Even G. Cuvier, the leading anatomist of the beginning 19th century, ignored them when he laid down his hypothesis resting on no real proof that bats located objects by means of their sense of touch. Their pressure sensitive frail wings were supposed to register the accumulated air that the animal should produce when approaching an obstacle.

About 130 years later mammal experts and physiologists attached the problem again and in a more profound way. In doing so they were able to make use of a substantially improved investigation technique, and the results were correspondingly good. What Hartridge had assumed in 1920 could now be actually demonstrated by Pierce, Griffin and Galambos, and independently of them, by Dijkgraaf (1943), i.e. that the flitting bats, active in the night, send out ultrasonic sounds and use the returning echoes to perceive obstacles and in this original way to make an "image" of their environment. The concept coined for this phenomenon by Griffin in 1944 was "echolocation". Indeed, it was conspicuous that in flying the animals kept their little mouths slightly open and gave off remarkable ticking and rattling sounds at brief intervals. Of course, if the bat was just about to fly off, or if you came closer to it, the same thing happened. One could assume that they were scanning the air space. Dijkgraaf now literally hung muzzles on the mouths of his tender-skinned charges. These made it possible to stop their sending out sounds. When this took effect the flying animals bounced on objects disposed in the experimenting room and on its walls. Likewise an aversion to flying developed right from the moment their ear canals were plugged.

120 Six faces of different bat species.
Nose leaves and big ears serve location and catching of prey as acoustic transmitting and receiving apparatus.
After M. Eisentraut.

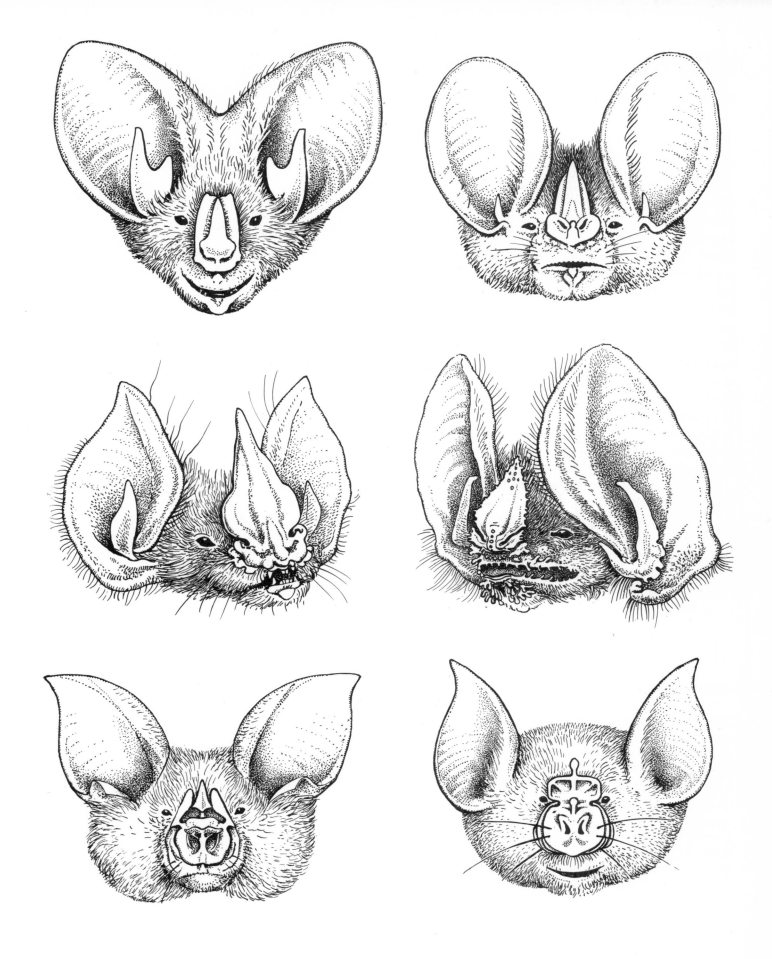

The researcher now logically formulated his finding that the orientation of bats was based on the perception of a special acoustic orientation sound which was sent out from the bat's mouth in a forward direction and was reflected by objects.

About the same time the character of these orientation sounds was explained by Griffin and Galambos: they were brief ultrasonic impulses whose frequency lay between 30,000 and 100,000 hertz. There is the following objection that might be raised: the ear of an adult human being can hear acoustic vibrations that lie between 16 hertz and a maximum of 20 kilocycles per second. Acoustic vibrations lying outside (ultra) this threshold of audibility are referred to as ultrasonic sound. The ticking and rattling sounds that are still only just audible for our ear are, as Eisentraut believes, to be conceived as additional noises by which the high-frequency orientation sounds of bats are accompanied.

In spite of similarities of the sound patterns in comparable situations, research into orientation sounds in different species as well as families established major differences as to frequency characteristics, duration and intensity. Frequency-modulated and constant-frequency sounds with a modulated final part have been found. Intensities of the sound vary greatly according to species: bats whose prey is small locate "loudly", while those with a larger prey do so with a fainter sound.

Griffin and Novick being specifically interested in the seat of the generation of these cries were able to demonstrate, by removing the nerves supplying the laryngeal muscles and by differentiating potentials of the muscle action during sound production, that the crico-thyroid muscles spread special laryngeal membranes which cause the ultrasonic sounds, the number of the sounds being adjusted to the orientation behaviour that happens to be desirable. It lies between 10 and more than 100 per second.

The ear, too, is well adapted to the special task of ultrasonic orientation. Apart from being highly sensitive to supersonic vibrations it is characterized by an excellent directional space pattern. In order that the very weak echo of the strong orientation sound may be at all perceived, the ear is protected from the strong orientation sound by a brief reflecting contraction of the middle-ear muscles.

The orientation performances that have been observed are truly amazing. It was recorded by Gould that in the case of the free flying American little brown bats *(Myotis lucifugis)* in 2,369 seconds each three seconds on an average they were chasing a new insect and it was estimated that every second one got actually caught. From the increase in weight it could be calcutated in a laboratory experiment that the animals when provided with the corresponding food supply caught a tiny pomace fly approximately every fourth second.

In order to comprehend the orientation performance in a quantitative way, bats were trained to fly through obstacles with vertically and horizontally stretched wires of varying

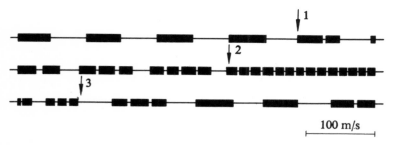

100 m/s

121 Duration and frequency of orientation sounds given out by a horseshoe bat during various phases of the flight when flying through a wire obstacle. At 1 there begins the approach phase, which at 2 passes into an intensive final phase until at 3 the obstacle is bypassed.
After H.-U. Schnitzler.

122 The two typically different frequency patterns of the orientation sounds of bats.
a—frequency modulated orientation sound of *Myotis lucifugus* (little brown bat),
b—constant-frequency orientation sounds with a frequency-modulated termination part of the horseshoe bat *(Rhinolophus ferrumequinum)*.
After H.-U. Schnitzler

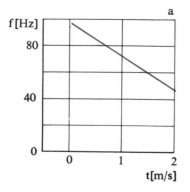

diameters and distances. Horseshoe bats were capable of registering wires of 0.08 millimetre diameter. The sound pattern of one *Euryala* horseshoe bat showed that in three out of eight flights it reacted to wires with a diameter of 0.05 millimetres.

Orientation performances of this order demonstrate that the animals are able to locate objects with exactitude and to identify them as to shape, size, and even material. In 1973 H.-U. Schnitzler pointed out that bats determine the direction of the returning echo and use this to measure the period between sound emission and echo and thus the distance of the object located. As for the information on these objects necessary for their identification, this would, however, have to be sought in the differences between the emitted orientation sounds and the reverting echoes. Now since the orientation objects are supposed to alter the relative intensity ratio of the individual frequency participations in the total echo while forming the echoes on the basis of their size, shape and their material, the bats would apparently seem to gather the information essential for recognizing objects by realizing and analyzing the echo structures characteristic of the respective objects of orientation which are also subject to alteration depending on their position and movement.

In view of the multiplicity of information contained in the echoes it is interesting to pose the question why it is that precisely ultrasonic sounds, and not such signals as are to be perceived by human beings as well, are emitted. In trying to answer this query the fact must primarily be taken into account that acoustic waves are only then reflected from an object with noticeable intensity when the dimensions of the object are not too small in relation to the wave-length. If, for instance, in the case of spherical or cylindrical objects the radius gets smaller than one-sixth of the wave-length (Schnitzler), then the echoes are of such low intensity that even bats, which are after all capable of still registering even a ten-thousandth part of the intensity of the orientation sound emitted, experience difficulties in perceiving them. Ultrasonic sounds with a frequency of 100 kilo-cycles per second have a wave-length of 3.3 millimetres, which is calculated from the relation $c = \lambda \cdot f$, where λ is the wave-length, c the velocity of sound (330 m/s in air) and f is the vibration frequency. Since the prey to be located consist mostly of smaller insects with dimensions of a few millimetres and even less, the conditions for the reflection of ultrasonic waves are favourable. For example orientation sounds with a frequency of 3 kilo-cycles per second at which the human ear possesses the greatest sensitivity of hearing would be unsuitable for bats' echo ranging purposes as the appropriate wave-length amounts to as much as 11 centimetres. Now it will become immediately easy to understand why there are changes in the mutual relations between intensities of the various vibration sections of a frequency-modulated orientation sound on its reflection from a small object of orientation, in fact, higher-frequency sections are reflected better than lower-frequency ones, which makes it possible to differentiate objects as to their size.

Distance determination demands an exact measuring of time, for e.g. the echo from a 50-centimetre distant flying object reaches the bat already in about 3 milliseconds. Experiments have proved that even distance differences of 2 to 4 centimetres are still distinguished by bats. This corresponds to a flight time of ultrasonic sounds lasting approximately 200 microseconds.

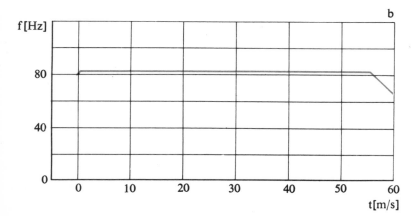

The relative speed between the flying hunter and its prey is calculated on the basis of the frequency alteration experienced by the orientation sound as a result of the Doppler effect. If the bat and the insect are moving towards each other there appears a frequency increase which grows with the increasing relative speed and lies in an order of about 1 kilo-cycle per second. Further pieces of information for a precise spatious localization are supplied by the very movable ears which, owing to their excellent directional characteristic, are capable of determining exactly the direction from which the echo is coming.

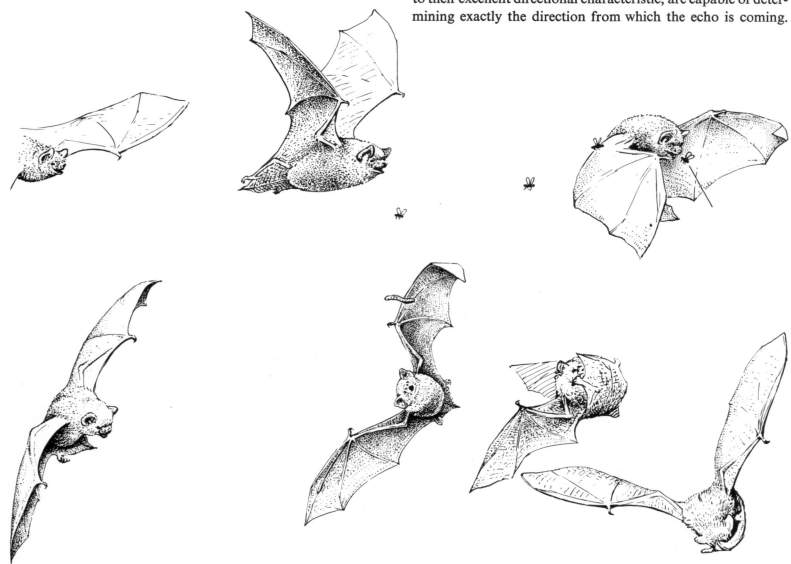

123 How a vespertilioned behaves when catching prey. Objects are a pomace fly and a thrown-up mealworm. After H.-U. Schnitzler.

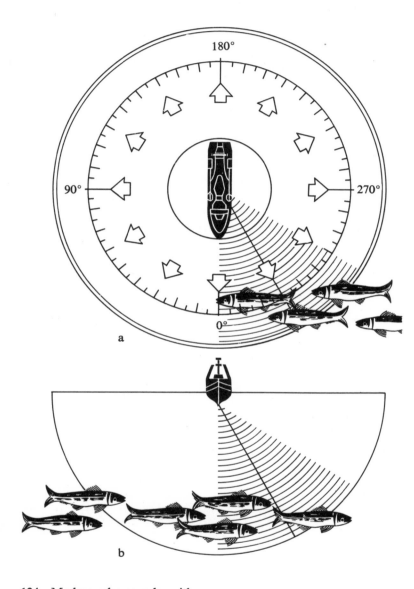

124 Modern echo sounder with which one can search for fish in the horizontal (a) as well as in a vertical (b) plane.

Incidentally, let it be mentioned here that ultrasonic sounds can be very sharply clustered and emitted with a substantially higher intensity than sounds in man's audible range.

Hunting bats cover about 5 to 10 metres per second. Thus the animals when flying against a 50-centimetre distant object have less than a tenth of a second for processing echo-communicated items of information in the central nervous system, for consequent orders to the muscular system, and hence finally for a purposeful reaction. When they approach it even closer, then the time intervals left over for these processes move in the range of hundredth parts of a second. When one considers with what skill these nightly flitting animals actually do catch about 50 per cent of all prey they locate, the amazing efficiency, and the reliability of the ultrasonic orientation system stand out with particular clarity.

Of course, it should be mentioned that "echolocation" fails when fog or mist sets in: then the orientation sounds get diffusely dispersed on the minute fog or mist drops and thus become useless to the animals.

When searching for ways in which man has applied ultrasonics for his own purposes, then the earliest application is perhaps the one used as a means of signalling propagation of sound waves in navigation. The echo depth sounding developed by A. Behm in 1912 in which the depth of the water under the keel is determined by the propagation time of acoustic waves from the vessel to the sea bottom, much the same as bats determine distance by means of ultrasonics, has been substantially improved. Under conditions of total silence—the applied frequencies lying far above those produced by the ship's machinery and other equipment—it has become possible to make from 7 to 15 measurements in a second so that the shallows can be sounded with certitude. Until the early thirties echolocation was being used exclusively for nautical and surveying measurement purposes, though as early as 1928 Behm had taken out a patent on his idea to use ultrasonic sounding for locating fish and fishing grounds. By implementing his idea fishing has become substantially more economical, since by means of echolocation the whereabouts and the expanse of a shoal of fish can be exactly determined, and hence the depth of the dragnet can be optimally adjusted. Echolocation also makes it possible to obtain information on the structure and constitution of the sea-bottom. From the manner and shape of the multiple echoes it can be perceived whether the ground is formed by rock, sand, clay or mud.

The echo sounder consists of a generator producing ultrasonic impulses, a sound transmitter and receiver, an amplifier for the echoes, an indicator and a recorder. The transmitter and likewise the receiver are mounted outside on the ship's bottom near the keel. The echoes from the sea-bottom are amplified and recorded by a recording instrument or a cathode-ray tube.

In this way the flight time of the sound impulses and thereby the water depth are recorded, for the distance that has been covered by the acoustic signal in all, i.e. from the ship and back again, can be determined from the flight time and the velocity of sound with the help of the relation $s = v \cdot t$. Velocity of sound in water amounts to about 1,500 m/s, that is more than a fourfold quantity of its value when travelling through the air.

Consequently, up to a few hundred metres of water depth it is possible to make a number of echo soundings per second. On the other hand in measuring deep sea rift valleys with abysses reaching down to 10 kilometres the impulses run as many as a few seconds so that a single measurement also requires several seconds to accomplish.

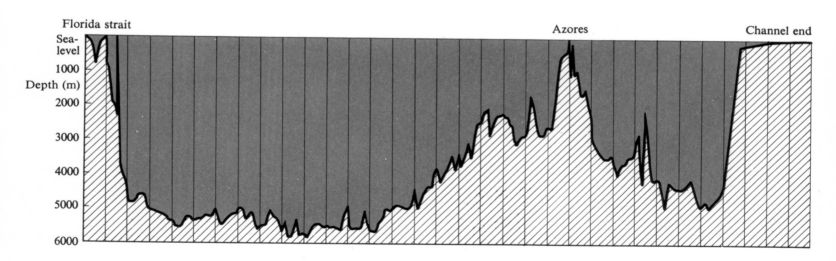

125 Sea bottom profile reproduced by means of echo sounding between the end of the English Channel and Florida. After L. Bergmann.

The Sharp Eye of Birds of Prey – Structural Peculiarities of the Retina

Stimulated by the echo direction-finding of bats technicians have developed instruments, based on the same orientation principle, for guiding the blind. Ultrasonic impulses are spread over a walking-stick and their echoes reflected from objects are converted into a tone within man's audible range. In this way the blind person can "hear" the obstacles obstructing his way.

The Doppler effect is used according to the principle of echolocation by means of ultrasonics also for measuring the speeds of vehicles. Of course, this process is limited by the relatively low speed of sound propagation, and thus loses its appeal to air travel in particular. In the latter field radio range-finding and laser direction-finding have been developed using electromagnetic waves as signals. These propagate in space at a velocity of 300,000 km/s and, unlike acoustic signals, are not tied down to a medium. It is with their help that distance measuring and location in air transport and space travel has become possible during which signals must traverse the ultra-high vacuum of space and, in addition, relative velocities of several 10 km/s may occur.

From times immemorial birds of prey have had an outstanding sharpness of sight ascribed to them. When one turns over the leaves of books on the physiology of senses concerned with birds this conception is found to prevail until quite recent times. Many an author supports his statements by the results of fieldwork in the sphere of ornithology and of histological findings dating from the early forties to the effect that in the retina of a buzzard eight times as many receptors per unit area are to be found as in man. Weighty dissenting voices were raised against this view in the sixties, and more recent researches have shown that the large birds of prey such as griffins, vultures, eagles, etc. are actually capable of giving better orientation performances with their eyes than man but, on the other hand, the latter are then again not so "overwhelming" to allow us to describe them as unique record feats.

None the less, the eye of a bird of prey does show certain peculiarities by comparison which enable the great dry-land sailers and carrion eaters to be particularly "keener of sight" than humans. How often can one read in books of travels that in tropical regions two or three dozen vultures descended on the carcasses of large fallen mammals within a short time though prior to that but very few were to be seen circling in the sky. Today we know that their success in searching for nourishment is due to their visual acuity and capacity of association. "Thus the individual vulture soaring in great heights is not dependent on himself and his good eyesight alone. He co-operates with neighbouring birds distributed in the air space, and with marker birds keeping near the ground." (cf. in A. B. Fischer, 1969) And it is always equally impressive to watch how purposefully e. g. a buzzard swings down from his elevated observation post amidst the village fields—perhaps a telephone mast that is 6 to 8 metres high—in order to kill a vole 50 or even 100 metres down to the side. It was indeed a dashing way in which such a hunting flight was carried out by a booted eagle *(Hieraaetus pennatus)* in Western Tien Shan; there one of the authors had the chance to admire one. The bird, one of the liberal buzzard size, was soaring high above the late summer scorched valley ground encompassed by gorges sloping down steeply over 400 to 500 metres. Suddenly, it began to "step down" as if descending a gigantic staircase. Rapid headlong falls of a falcon-like flight pattern alternated with a soft vibration flight in one spot, the eagle keeping its claws dangling, apparently to put its centre of gravity as low as possible to get a good hold of its prey. A rush, quick as lightning, down from the remaining 30 to 40 metres, and the impressive manoeuvre was over.

To clarify the concepts let it be noted that in the ensuing discourse "keenness of vision" signifies the spatial resolving power, in which a distinction must be made between selectivity and point definition. "The selectivity of sight serves the animals to recognize objects according to their shape; it provides them with genuine stereoscopic vision. The point definition, on the other hand, is of importance in cases where it is a question of discerning objects which stand out from their background by reason of contrast or motion without having to be identified as to their shape. The measure of the keenness of vision is given by the smallest optic angle under which a point or a single line can still just be observed, the 'Minimum visibile', or under which a number of equidistant lines can still be distinguished from one another, the 'Minimum separabile'." (cf. in A. B. Fischer)

Since the superiority of the eye of a bird of prey over that of a mammal proves ultimately to be a consequence of structural peculiarities of the retina it may be an advantage to give a brief description of the overall anatomy of this descendant of the brain; so that this superiority may be the easier to understand. Among the vertebrates living in the countryside it is birds whose eyes are relatively the largest. The keenness of sight is determined by the number of optic cells in the retina. When scanning the background of the eye with light from the opthalmoscope one can see this retina extended. It consists of several layers of nerve cells of which of course only the one turned towards the crystalline lens is sensitive to light. The "visual layer" is composed of an immense number of sense cells in the form of cones or rods. Depending on whether their bearer is active by day or by night, it is either cones or rods that predominate: the former make coloured perception possible, the latter serving non-coloured vision. In the fields of vision, the "areae", in the middle sections of the retina, the cones are very closely concentrated, and a particularly distinct, fixing sight is made possible by little visual pits (foveae) to be found in them. In addition to a central visual pit, birds of prey are equipped with a side visual pit as well, which enables them to fix their animals of prey with one eye and both eyes at once. This gives them the use of three fixing directions in the field of vision. The structure of both the foveae enables them not to "lose out of sight" even smaller or distant moving objects. "A thorough investigation of the shape of the foveae is due to Pumphrey (1948). This has shown that in view of the lens effect of its convex walls the deep funnel-shaped retina pit provides for a quick apprehension and recovery of moving objects as the image running in the fovea alters its speed jerkily. The 'distortion effect' of the fovea is all the greater, the steeper the funnel walls are." (cf. in H. Oehme, 1964) The reader is sure to have noticed something analogous to this when lit by the sun from the side while walking along a house front. His shadow image which until then had accompanied him on the house wall with equal speed shot, all of a sudden, like a bolt into the new direction as he turned the corner. The abrupt speed alteration, the "jerk in the hitherto uniform motion makes for easier observation as well as fixation" (after R. Berndt and W. Meise, 1959).

In large birds of prey spatial resolving power, the keenness of vision, is comparatively larger in the extrafoveal retina as well, i.e. the retina situated outside the little visual pits for, according to Fischer, the large participation of cones in the periphery allows birds in general the utilization of colour values over the entire field of vision; moreover, their sharpness of sight does not decline on the periphery to such an extent as in man. As a result, eagles and vultures have simultaneously a panoramatic field of vision, and the picture is completely coloured and sharp. For some species of vultures Fischer has calculated that given a keenness of vision of 13.3 to 17.2 seconds of an arc visual objects of as little a size as 23.5 to 30.4 centimetres can be discerned from the height of 3,650 metres. "Migrating herds of antelopes, lions tearing their prey or even crows at a carrion can be followed with the eyes."

Sensibility to motion enables the bird to perceive up to about 150 successive images per second. In a cinema up to 24 images per second merge before our eyes depending on the brightness. However, a peregrine falcon, even when falling upon its prey in a madly rushing dive, always preserves a kind of slow-motion survey of its environment; its eyes are well capable of resolving the rapid succession of pictures as though in slow motion.

Let us sum up the set of problems with the moderate words of H. Oehme: "Presumably, just as in the primates, the extra-foveate retina also plays the part of a search organ; the fixing of an object is then performed with the fovea but the bird will have its attention drawn even to a tiny motion which we, with our own retina functioning with less fineness in this respect, are unable to register. The reason is that even outside the fixation points the tie lines of the inner nuclear layer of the bird's retina (the powerfully developed second layer of neurons, see drawing—authors' note) in combination with the much more numerous cones constitute an organ which is far better suited for this function."

126 Partly schematic sections through the foveae (left: in the centre, right: on the sides) of two common buzzards. Shaded: Layer of rods and cones. Black: Layers of cell nuclei (outer and inner nuclear layers, layer of ganglion cells).
After H. Oehme.

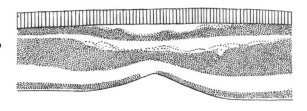
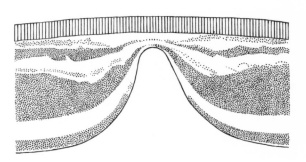
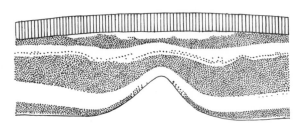
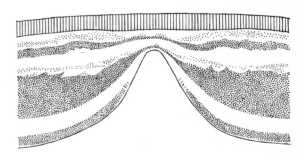

Pit Vipers Search for Prey with Infrared "Eyes"

The fact that when the surface of the body hit by radiant heat is perceived by the thermal sense-organs on the skin is a truism which would hardly need enlarging upon. If we let ourselves be irradiated by a heat source then the latter can also as a rule be located even when one does not look out to see where it is situated. Of course—compared with the small radiation energy that is enough for a source of light to be observed with the eye the amount of the former must be much more substantial to enable man to make out heat radiation.

Not a few animals are capable of registering the ambient temperature in a differentiated way. For many it has even become a decisive factor in building an adequate ecological niche for themselves. At least, according to the present state of knowledge—the perfection of this principle of taking in radiated energy and using it for an entirely definite life-preserving orientation performance is represented by two groups of snakes: the pythons and the pit vipers, the latter including the well-known rattlesnakes of the New World. They are equipped with highly sensitive "thermometers" which are at the same time outstanding radar direction-finding instruments and serve the location of prey. At either side of the snake's head, a little beneath the eye-nostril level, there is a little pit (hence the name of the species) embedded like a steep-walled funnel. The edge of the pit appears to be corded up, beneath it the funnel extends into a bulge. The ground of the pit is spanned by an extremely fine membrane "containing an entanglement of tender blood vessels and a dense nervous network" (cf. in P. Raths and G.-A. Biewald, 1970). There is still some air space behind these tiny membranes—they have a "thickness" of about 15 μm. This structure has a very slight thermal capacity and thus prevents the membrane from storing up heat, though it does receive for a brief space of time a dose of perceivable heating, the radiated heat which is particularly strongly absorbed being that with wave-lengths between 1.5 and 15 μm (frequencies of about 10^{14} to 10^{13} Hz).

It may be to the purpose to include here a few remarks about the way the physical nature of calorific rays had been discovered: Sir William Herschel (1738–1822), Astronomer Royal of King George III of England, the man who among other things discovered the planet Uranus, established in 1800 while observing the sun that the spectrum of the light visible to us (wave-length 0.4 to 0.75 μm) is accompanied by an invisible heat radiation outside the red range. He traced the solar spectrum—produced by means of a glass prism—with the aid of a quicksilver thermometer and found a particularly strong heating close to the still visible red range. The name given by Herschel to this unknown, invisible radiator was "dark heat".

It has since become generally known that this radiated heat beyond the red range—hence also called ultrared or infrared (IR)—consists of the same electromagnetic waves as e.g. X-ray radiation, visible light or radio waves. All of these propagate with the velocity of light, the only difference being in the wave-length. It is also well known that each body whose temperature lies over the absolute zero point (—273 °C) sends out heat radiation whose wave-length decreases with the growing temperature (if a body is sufficiently strongly heated, it radiates visible light in the manner of tungsten coils in an electric incandescent lamp). Men and animals, plants and buildings emit radiation in the region of around 10 μm, the whole IR-range lying between the 1 μm and 1 mm wave-length. And it is just at 10 μm that the "ultra-thermometer" operates which enables its scaly bearer to follow the heat traces of its small animals of prey lying in the ultrared range, thus to track its nourishment even in total darkness. After all, it occasions no surprise to find just snakes as very "cool" animals with alternating temperature, and not perhaps the much more highly developed warm-blooded animals, endowed with sense-organs sensitive to ultrared light; indeed, in the latter their own heat radiation would block such organs from receiving outside information.

The way these reptiles acquire their prey is clearly illustrated by the physiologists P. Raths and G.-A. Biewald of Halle: "While stalking in the dark they crawl slowly around, searching the bottom or the branches of a bush for something that is warmer or cooler than the environment. A difference of a mere two-thousandth of a degree is enough to alert the serpent. Thus it is able to perceive other animals already at a one-metre distance. Warm-blooded animals excite the minute nerve-fibres of the membrane, while the cold-blooded ones inhibit them. The pit organs are excellent direction-finding devices. With their steep walls they operate like a concave reflector by which all heat rays are collected and projected directly on to the focus, the fine filmy skin, so that the prey can be exactly located. Some of these vipers stick their venom fangs in the victim and then let it go. Dying it escapes into a hiding place but the reptile persistently follows its temperature track until the dead animal is found."

The drawing on this page illustrates that each of the pit organs is spread over a recording cone. In this range the heat radiation is perceived. Close to the muzzle tip the pit fields overlap, thus giving rise to a binocular field of "vision" in which a radiation source is discerned spatially; both the infrared "eyes" pick out warm-blooded animals of prey as well as those with alternating temperatures from their mostly constantly tempered background. The dense nervous network of the pit membrane responds to every increase in the ambient temperature near the head by substantially multiplying its impulse density. The required excitation threshold lies as low as under 0.005 degree of temperature difference.

127 Infra-red orientation in a pit viper. In the indicating cones of the little pits prey is exactly located.

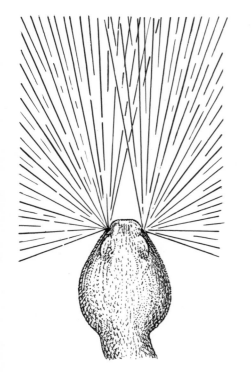

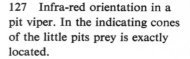

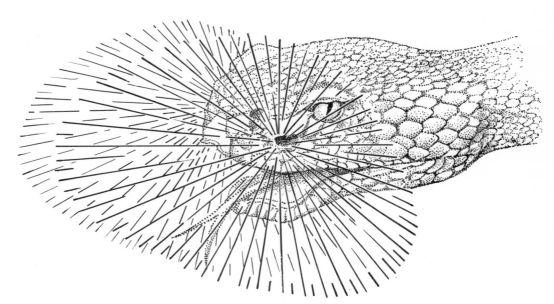

Moreover, it is interesting to note that a markedly slow change in temperature—under laboratory conditions—produced but a slight change in the frequency of the nervous impulses in the pit organs. On the other hand, jumps in temperature (of 0.2 degree per second) increased the impulse frequency for a time by as much as more than a half. This process is inherent in all sense-organs. It is closely connected with the adaptation of the sense cells—also illustrated in the introductory chapter. After an initially high impulse density as sequel of an initially higher stimulus which subsequently remains constant, this adaptation results in substantially reducing the number of nervous impulses per second. This reaction on the part of the receptors is very "reasonable", for in all orientation phenomena in the animal kingdom—whether during prey acquisition or while escaping from a pursuer—it is primarily the temporally and spatially changing outside stimuli that really matter.

Less than 200 years have elapsed since the discovery of the infrared rays. Their application in science and technology received a powerful impetus especially after the Second World War. Thermal rays obey the laws of optics in the same way as visible light does. Of course, only few natural materials are transparent enough in a sufficiently large range of wave-lengths (e.g. rock salt). Since about 1930 there have been many synthetic cristals at our command which are permeable deep into the infrared range. From such materials lenses and prisms can be made for reproducing optical images and used for making invisible "heat images" of objects, living beings, or landscapes. To make these "heat images" visible requires image converters. IR-sensitive detectors take up the thermal rays and convert them into signals by means of which a visible image can be produced.

An example of a detector and an image converter in one is the infrared sensitive film, which can of course work only at a relatively short-wave radiation (about 1 μm) as otherwise it would turn black through the long-wave fluorescent radiation of the camera itself. The development of detectors passed from the quicksilver thermometer over thermic elements (Nobili, 1829)—with these one was able to measure a man's thermal radiation at a distance of 3 metres—, the bolometer (Langley, 1880)—this made it possible to prove the heat of a cow at a distance of 400 metres—up to the modern highly sensitive photon detectors designed on semi-conductor basis.

The military uses of the IR-radiation are legion. Soldiers could be spotted in the dark at many hundred metres by their heat radiation. Lorries, ships, aircraft—nothing escapes the "infrared eyes". In the case of aircraft the heat radiation can still be detected across many tens of kilometres.—After 1955 a tempestuous development was initiated in the civilian sphere as well. As late as 1960 a ten-minute exposure was necessary to produce a serviceable thermal image. Soon afterwards, however, entirely new opportunities offered through the use of a new scanning system with which the IR-picture could be scanned similarly as in television though in an optomechanical way. A Swedish firm builds devices with a picture frequency of 16 hertz. The technique of making pictures by means of invisible thermal radiation is called thermography. In the last few years it has conquered a wide field of application. In the first place, diagnostics in medicine, particularly for an early recognition of a mammary carcinoma (diseased and sound tissue spheres emit differently in the IR-range). Furthermore, it is possible to find out reliably and locate heat pollution of waters, heat leakages in buildings and industrial premises (e.g. blast furnaces or heating pipes under roadways), hot points in generating plants and many others. With the IR-microscope it is possible even in microranges—e.g. on integrated solid state circuits—to detect local overheatings and thus to determine and eliminate the causes of deficiency of electronic components.

Among the many tropical families of fish "low-voltage fish" have only been known for the last two decades or so. On the other hand, "high-voltage fish" had been known since antiquity and there was exact knowledge concerning the power of the electric blows dealt by the electric catfish (genus *Malapterurus*) in order to benumb his prey and to defend itself against its enemies. However, the nature of this phenomenon was not at all clear; the knowledge of the existence of electricity was to mature only two centuries later.

"High-voltage fish" are encountered in both fresh water and sea water, their classical representatives being the already mentioned electric catfish, the electric eel *(Electrophorus electricus)* and the electric ray from the genus *Torpedo* and *Raja*.

"Low-voltage fish" outnumber by far the former, but are mostly inferior to them as regards the voltages produced. As will be seen this difference in the specific performance capacity also underlies an essential difference in the function of the electric organs. These are, except for the electric catfish, converted skeleton muscles to be found in various places of the body; thus electric eels have them in nearly the whole caudal region, many electric rays on either side of the gills. In the electric catfish the electric organ envelops the body like a cloak and it most probably originates from modified skin glands. Consequently, the volume of these batteries is considerable and in the case of the electric eel can make up more than a half of its body weight.

These electric cords are constructed like bundles out of columns which in turn consist of jellylike plates (the modified muscle fibres) piled up on each other in layers. There is always one nerve fibre that grows into each plate in the course of embryo development, hence individual elements are always unilaterally innervated. Each plate is to be conceived as a simple galvanic cell. In the columns these basic units are coupled in series, and a contemporary discharge of all occurs as soon as the appropriate order from the fish's brain gives a nervous impulse to the whole organ. According to the number of the respectively constructed plates the individual voltages are added together and it is easy to imagine that a prey fish of an

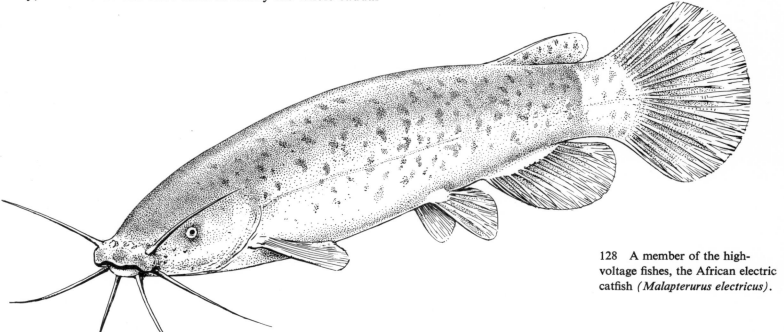

128 A member of the high-voltage fishes, the African electric catfish *(Malapterurus electricus)*.

electric eel instantaneously becomes incapable of manoeuvring when hit by the first electric volley of its pursuer. In any case, 300 to 800 volts have been "delivered": the 5,000 to 6,000 plates of the hunter supply individual voltages of 0.06 to 0.12 volt each. Even a man would immediately be stunned by their total sum. Of course, the total current of the electric blow flows only

a short time, i.e. a few tenths of a second (depending on the species the individual discharge lasts from 2 to 6 milliseconds). The voltage is measured after the number of plates connected in series. The power of the voltage increases with the number of columns as these batteries are connected in parallel. In this way 60 to 120 amperes are produced by the approximately 500

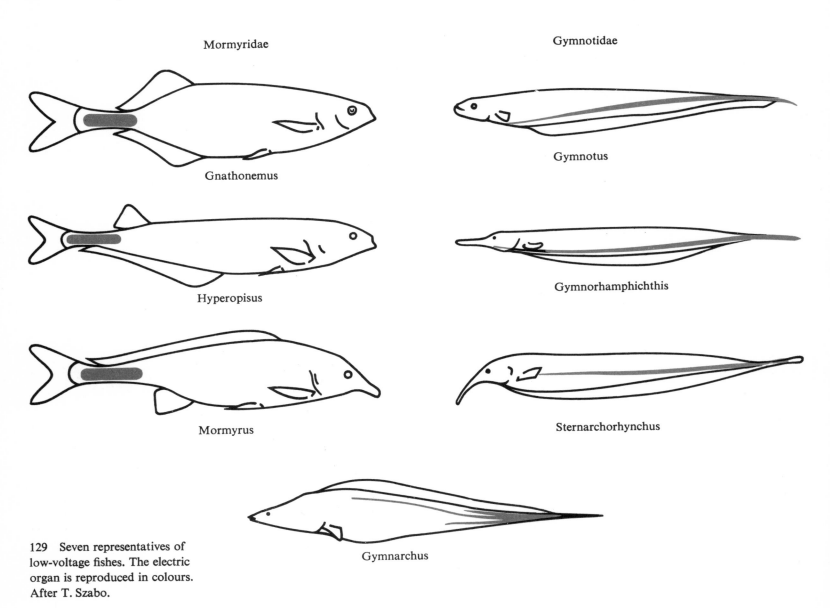

Mormyridae

Gnathonemus

Hyperopisus

Mormyrus

Gymnotidae

Gymnotus

Gymnorhamphichthis

Sternarchorhynchus

Gymnarchus

129   Seven representatives of low-voltage fishes. The electric organ is reproduced in colours. After T. Szabo.

columns of big electric rays of the genus *Torpedo*. The electric eel with about 70 batteries manages no more than 1 ampere. As to voltage, 300 volts have been quoted for the electric catfish, for the various electric rays it ranges from 20 to 200 volts.

The many species of "low-voltage fish" use their electric organ for navigation. The few volts produced are not enough for catching prey and for defence against enemies. For the sake of completeness, however, one should mention that the "high-voltage" electric eel is also equipped, likewise on the edge, with a small electric organ by whose constantly emitted weak impulses it is enabled to locate animals of prey, obstacles and competing representatives of its own species in its gloomy environment. "Obviously, the *Electrophorus* is able to perceive fluctuations in its own electric field, as produced when the latter is traversed by other organisms, and the fields of its fellows of the species." (P. Raths and G.-A. Biewald, 1970)

In "electric" fishes a distinction is made between active and passive location. In the first case, the fish by means of its electric organ in a quick impulse repetition (several 10 up to 1,000 hertz)—builds up a field about itself which is, however, effective for location only in a narrow sphere of four to five centimetres distance from the fish (T. Szabo, 1974). The impulse control station is situated in the medulla oblongata. Electric receptors in the skin excited by the fish's own field sensitively record field deformations in the animal's vicinity brought about by foreign matter or other living beings. But various electric receptors are also capable of registering fields of "electric transmitters"—other electric fishes, direct currents in running waters, etc. This is termed "passive location". "The receptors participating in this location mechanism mainly serve communication between members of the species, an important role being played by discharge frequency." (T. Szabo, 1974) The threshold of response of receptors for active location in the close-up area lies relatively high. Moreover this location mechanism is protected from outside stimuli by characteristic discharge patterns.

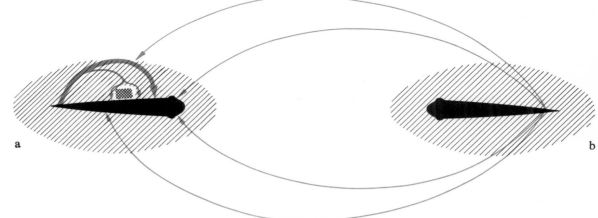

130 Active and passive location mechanism of electric fishes.
a—in the shaded antenna-proximity zone the fish locates actively by means of the electric field set up by itself. In addition, the fish (a) is in a position to perceive in a passive way electric fields of other fishes (b) and to locate exactly such natural "transmitters".
After T. Szabo.

The Honeybee—Solar Compass, Interior Clock and Dancing Language

Surprisingly enough, not only the mere existence of foreign bodies in the fish's close-up area can be located electrically. The fish is able to differentiate objects as to their dimensions, material constitution, position, motion and direction of motion. It is no exaggeration to say that these fishes have a perfect electric survey of their closer environment. Objects of minutest dimensions do not escape the receptors even in places where the lateral line organ and the eyes can no longer perceive anything, e.g. in turbid, night-dark or torrential waters.

The electric location mechanism differs fundamentally from the ultrasonic direction-finding of bats and dolphins. In the case of both, the actively emitted sound waves must first be reflected by the object to be located, and the reflected wave sections are then utilized as orientation stimuli. It also differs from radar, however, where the reflected section, though of electromagnetic waves which propagate with the velocity of light, is also used for locating objects. The specific stimulus for electric receptors, namely the field built up in the water, though propagating in a similar way as radar electromagnetic waves, i.e. with the velocity of light, always reaches the sense-organ— even without foreign matter—and is only modified by field distortions. It is these field distortions that contain the required information for orientation.

The bee's industry is proverbial. A fact known from school is that these active little animals while searching so indefatigably for food are at the same time of primary importance for cross-pollination and hence for the yield of many crops. The terms of the society of bees or bee colony, the queen, drones and workers, are generally known, and not a few of us have painful memories of having been stung by a bee when she felt herself endangered. But this is about as far as the knowledge of most of us goes.

For thousands of years now man has taken possession of the coveted products—honey and wax—of these interesting social insects. Though bees have now been domesticated for a long time, it was not until a few decades ago that their excellent orientation sense, or their communication with the other members of the species, began to be explored. It was particularly the famous behaviourist, the Nobel Prize winner Karl von Frisch who, with a team of enthusiastic scientists and assistants, in the course of about the last fifty years has succeeded in lifting the veil from many secrets of the bees and tracking the almost unbelievable capabilities and sense capacities of these animals.

Bees make scout flights from their hive to a nectar source which is often kilometres away. Whereupon—fully laden with pollen and nectar—they return in the shortest way—"taking a bee line" as the saying goes—to their abode. Shortly after the return of these "scouts", the so-called scout bees, many of their mates leave the stock and direct their way—unaccompanied by any female scout—purposefully to the feeding source. How did the scout bee on its quest flight take its bearings and even calculate the shortest way, the "bee-line", back to the hive? Yet even more enigmatic appears the fact that she has obviously communicated to its bee mates such data as are necessary for the finding of the source, i.e. direction of flight and distance. Today we can say that all this has been thoroughly clarified. For its orientation the bee is equipped with a compass and an interior clock, as well as a distance-measuring equipment. During phylogenesis a unique kind of "mute language" has evolved for communication with other female members of the species—the dancing language. Compass, clock, distance-measuring equipment, language—many a reader may feel it an exaggeration to apply these concepts to an unpretentious in-

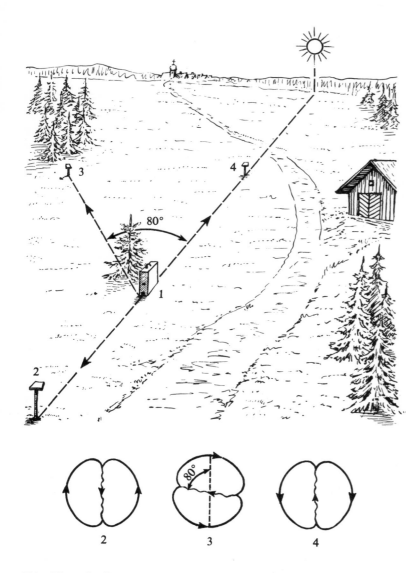

131   Three feeding sources distributed in different directions around a bee hive and below the corresponding waggling dance figures.
After K. v. Frisch.

sect, and yet they hit the nail on the head, or the heart of the matter, as one might say. The sun serves as a compass for the bees. The angle between the direction of flight (longitudinal axis of the body) and an imagined perpendicular from the head to the sun, measured with the faceted eye, is communicated to the bee's mates as something like marching orders towards the goal. However, everyone knows that the position of the sun constantly changes. On its apparent orbit round the earth the sun covers an arc about 15 degrees in an hour. Flights of bees may quite easily occur within one hour's range, and during this time the azimuth of the sun's position alters considerably. In addition, the scouts in their dark hive even after hours—without having left the abode in the meantime—currently inform the other working bees about the new correct angle between the direction of flight to be set and the actual position of the sun. Therefore they must be capable of calculating the position of the sun at each particular moment. Yet this is only possible if they can determine the time that has elapsed from the moment they measured the angle themselves to the moment when they communicate it to the fellows of their species. But time measurements require a clock and such a clock is carried by bees in their insides. They are said to have an endogenous clock. This time-measuring instrument also functions under exclusion of all environmental factors, concerned with the periods of a day, but is set at local time. In the course of the genesis it was, however, programmed through circadian (day-periodical) stimuli. Which biological processes actually function as time indicators is so far unknown.

Nevertheless, bees are not the only animals with an interior clock. For instance, birds of passage also need exact time measuring to determine their course. So have explorers of former times observed an outstanding sense of time in primitive races. With advancing civilization and with shift labour and other ways of life contravening the natural rhythm of the day this sense has gradually been suppressed in man. Fortunately, he has succeeded in devising artificial time-measuring instruments (chronometers) for himself ranging from the sun-dial to the atomic clock, since even for him the knowledge of time plays the decisive role in many orientation problems on the revolving globe. For instance, almost until the middle of the

18th century it had been extraordinarily difficult to determine the position of ships on the high seas. A position on the earth's surface is given by the geographical latitude and geographical longitude. The former can easily be calculated from the height of the stars over the horizon, while the meridian, on the other hand, cannot be established in this way. To measure the angle according to constellations by means of a sextant is important but an exact determination of time is equally essential. The whole thing proceeds as follows: at the ship's departure the chronometer is set at the time of the departure meridian (the zero meridian passes through Greenwich). If the position is to be determined the navigator uses the sextant to measure the altitude of the sun over the horizon and from this he determines the local time, let us say 12.00 o'clock. If the chronometer then shows 1 o'clock the vessel is 15 degrees away from the departure meridian. Two minutes of time correspond to 0.5 degree, that is a distance of 100 kilometres on the equator. Consequently, a clock keeping very precise time was the basic prerequisite for knowing one's exact position. It was at the instigation of the great Newton that the British Queen offered a prize of 20,000 pounds sterling for the discovery of a method that would allow to determine the geographical longitude daily during a voyage from England to the West Indies with a precision of 0.5 degree. It was John Harrison (1693–1776)—he is acknowledged as the inventor of the ship's chronometer—who thereupon built a clock with which this requirement could be fulfilled. His "Timekeeper No. 4", which was tested in 1761 during a voyage from Portsmouth to Jamaica on the "Deptford", showed a difference of a mere 5 seconds after a sailing time of 161 days. Today there are even much more precise chronometers in existence, but their importance for determining position and course has been lost in the meantime. Every ship can find out its position from the point of intersection of two lines made up by signals from radio beacons.

But let us now return from our little excursion into navigation back to the bee.—Compass and chronometer have been discussed, but in what way are distances, or better still, the length of path covered really measured? Exact investigations have shown that it is "fuel" consumption (sugar) and not perhaps duration of flight, or an optical control of the ground below, that is utilized for determining the distance. For if the bee has flown against the wind, then the distance communicated to her mates is too long. On the contrary, if she flew with the wind "in the back" then the distance given is understated.

Having referred to communication with the bee's mates on several occasions we feel it an obligation to elucidate the "language" of these useful insects in greater detail. Both larynx and lungs are missing; consequently the language is not "spoken" in the human sense of the term. The necessary orientation data are simply communicated to the other inmates of the hive in the form of a dance. What is the way this happens? Let us assume that a scout bee has come back from its foraging flight, and may find in her hive a horizontal plane from which she can see the sun. On this plane she starts dancing describing approximately the figure of a horizontal eight (see Fig. 142).

Soon she is joined by other bees in this dance referred to as "the dance of the foraging bee" or "waggling dance". The direction to the feeding source is given directly by the straight line between the two "zeroes" of the eight. On the other hand, the distance of the target must be deduced by the following bees from the rapidity of the dance. With increasing distance the scout bee slows down the rapidity regularly. There is an unequivocal and established relationship (Fig. 143) between the number of the dance figures per minute and the actual distance. The connection between dance rapidity and distance could still be proved for paths of as many as 11 kilometres. Information about the nature of the tracked feeding source is given by the adhering scent, and the eagerness of the insect's dance gives some hint as to its abundance. Moreover, while many of the lower-organized animals possess only one chemical sense, the bee has two: she can smell as well as taste.

133 a—Bats flitting through the air space when catching their prey by night.
b—Mouse-eared bat *(Plecotus spec.)* with overdimensionally developed external ears.
c—X-ray photograph of a Micro-chiroptera which especially demonstrates the transformation of the forelimbs into a flying organ.
d—Small horseshoe bat *(Rhinolophus hipposideros)*.

a

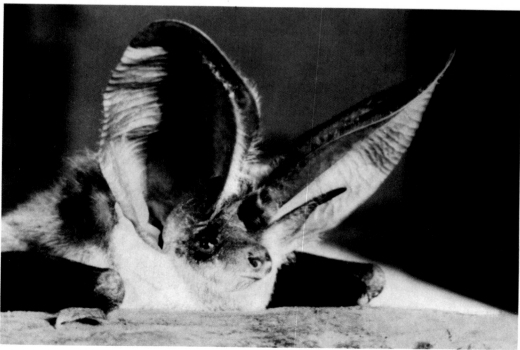

b

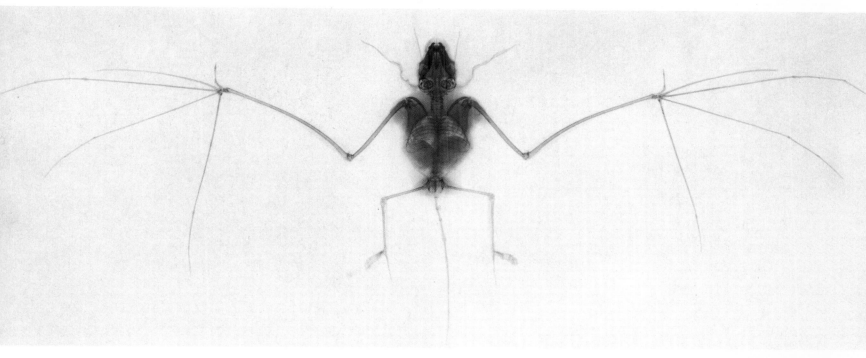

c

d

134 The South-American electric eel *(Electrophorus electricus)* deals out particularly heavy electric shocks.

135 The electric organs of the electric ray fish are kidney-shaped structures on either side of the front axis of the body.

136 Solar compass, interior clock, capability of measuring distances and dancing language are responsible for an almost incredible orientation and communication performance of the honeybee.

137 By means of the waggling dance (dance of the foraging bee) bees convey the news about a nectar or pollen source.

138 Thick panties of pollen "decorate" the legs of this busy collector.

139 Movable apiary in a lucerne field.

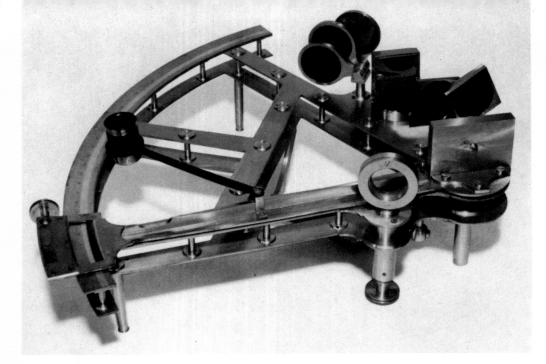

140 Hadley's sextant. With this instrument for angular measurements (invented in 1731) the parallel of latitude of a ship's position can be determined.

141 Nautical chronometer of Parkinson and Frodsham, London, about 1835. The chronometer is suspended on gymbals in a wooden box for protection against the ship's rolling movements.

If the bee is forced to dance in a dark hive on the vertical honeycomb surface without a sight of the sun, then she transposes the angle to the sun into an angle to the perpendicular. In this she is aided by the sense-organs of the gravity orientation on the articulations of her body. These data are at once cor-

rectly understood by mates of her species and the latter having taken flight set the indicated angle towards the sun.

It is a truism to say that the sun is not always shining. Is not the bee's solar compass bound to fail when the daylight constellation is obscured by clouds? Decidedly not, for a little

142 Circular dance (left) and waggling dance of the honeybee. By means of the circular dance the scout bee informs her hive mates about feeding sources to be found within a distance of less

than 100 m from the hive. The exact distance is indicated by the rapidity of the dance.
If the distance of the feeding source exceeds that of 100 m the bee is capable of indicating even

the direction by means of the waggling dance (dance of the foraging bee)—by giving its fellows the angle between the flight direction to be taken and a perpendicular to the sun.

1 Arrow pointing towards feeding source,
S Arrow towards the sun,
α Angle.
After W. Nachtigall.

piece of blue sky is enough for these animals to find their bearings, nor is a thin compact layer of clouds enough to jeopardize their sense of direction. What helps them to keep on is a capacity not possessed by man but one which is widely spread in the insect world—i.e. the ability to determine the direction of oscillation of polarized light. The sun's position and the polarization pattern are unequivocally interrelated, and the bee is capable of converting one into the other. Consequently, orientation after the solar compass does not fail until the sky is completely overcast with thick compact clouds. Yet even then, at least in the vicinity of the hive, the experienced bee is not entirely helpless; she takes her bearings from familiar landmarks (trees, bushes, buildings, rocks, etc.).

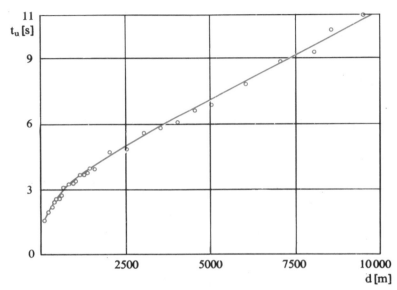

143   The connection between the period of revolution t_u, in which one dancing figure is run through by the scout bee, and the distance d of the feeding source. After K. v. Frisch.

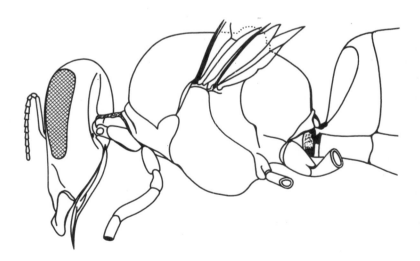

144   Sense-organs of the sense of gravity in a honeybee at the articulation of the thorax with the head and the hind part of the body. With the aid of these sense-organs the bee, while dancing in a dark hive on the perpendicular honeycomb surface, is able to convert the angle to the sun into an angle to the perpendicular. After K. v. Frisch.

Bird migration—what magic the word holds! Of course, to a town-dweller, and such are many of us these days, this kind of apostrophe may appear rather an idealizing one. For has he ever had a chance in autumn of an encounter with flocks of birds migrating in a narrow long formation? And even a countryman will mostly not take for migratory birds at all the many birds passing by on a broad front in search of their hibernating quarters in the south, often singly or in small groups, and flying none too high at that. The way they fly is too dispersed, too inconspicuous to make you turn your head after them. Perhaps there is a chance of this happening in October or November when you suddenly hear the striking "tsi-e-eh" cry of migrating thrushes.

And yet there are places on the migration routes where birds such as cranes or wild geese in their thousands break their journey in order to rest or sleep until cold snaps make them seek a more distant refuge in warmer climates. Such spots are generally tiny remainders of nature or nature-like landscape surviving amidst our spreading civilized surroundings, nearly always remote, not infrequently to be reached only at a considerable expense of time. It is this that has protected them from continuous disturbance by man, and, as a rule, it is quite impossible to find their whereabouts.

However, the man who knows and, carefully concealed, expects the evening invasion of Nordic cranes and complete military hosts of white-fronted geese and bean geese is looking forward to a spectacle of nature which makes him recede into the background on the scene in quite an unwonted way, turned into an insignificant marginal figure in view of the overwhelming copiousness of birds arrived from distant lands which fully dominates the scene captivating both the eye and the ear. Even today a man may now and then experience what was written by Johann Friedrich Naumann, the renowned German ornithologist in the early 19th century, about migrating bean geese.

"Having eaten their fill and tired of picking the corn, many a host stands even before sunset on its feeding place in dumb inactivity, partly with stretched out necks, partly squatting, to await the dusk in all stillness; only now it rises and hurries towards the water to reach it a good half an hour before night actually falls. As all detachments of such an army that generally assembles in this place arrive there within half an hour, and with a mighty noise, with every crowd looking for its own little spot, and as it takes them an almost equally long time to find it, or force themselves in between the others, all this seeking, selecting, invading, rising and repeated incursions occur amidst indescribable general confusion, a truly deafening noise. This is followed by sipping and cackling, then comes bathing until it has become completely dark, whereupon the noise gradually subsides, and finally dead silence falls with everyone abandoning himself to sleep until dawn breaks once again. At the very crack of dawn a still murmur rises among the birds. However, once the rising sun can be perceived still quite below the horizon, the multitude rises with uproarious cries, and in as many detachments as it had arrived, flies away to the field again, the very same field where they have found their feeding all those long years."

The last sentence already intimates the enigmatic phenomenon as it has appeared for centuries past: how do the migrating birds manage to find their destination and their way to it. Let us say in advance: there are many questions of this kind to which even in our days no—or at least no satisfactory—answers have yet been given. To be sure, many a sophisticated experiment carried out in the last two or three decades has established beyond doubt that birds possess various compasses, which clearly indicates the existence of a real navigation process—which, according to K. Schmidt-Koenig (1970) is the ability to maintain or establish a reference to a certain destination without the use of well-known landmarks. However, the fact should not be overlooked that what compasses actually indicate are not destinations but directions. By establishing orientation with the aid of a compass we have made but the first step on the road to explaining the navigation process itself. Up to the present there is hardly any knowledge at all of the sense-physiological basis underlying this phenomenon.

145 Migration routes of the European white stork *(Ciconia ciconia)*. The Central European migration boundary (right-hand points) clearly separates a small West front from the narrow East front, richer in the number of birds. After E. Schüz.

146 Unparallelled non-stop flyer: The golden plover of Eastern Siberia migrates (left) over a distance of about 3,300 kilometres from the Bering Strait to its winter quarters on Hawaii and Marquesas.

The American golden plover carries out (right) a loop passage between the arctic Canada and the pampas of Central South America. In doing so it covers 4,500 kilometres across the open sea. After F. Salomonsen.

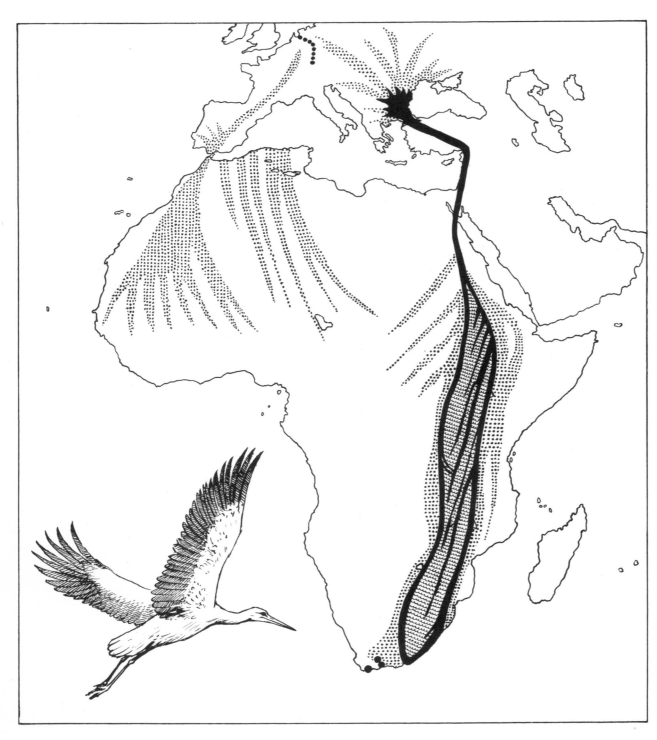

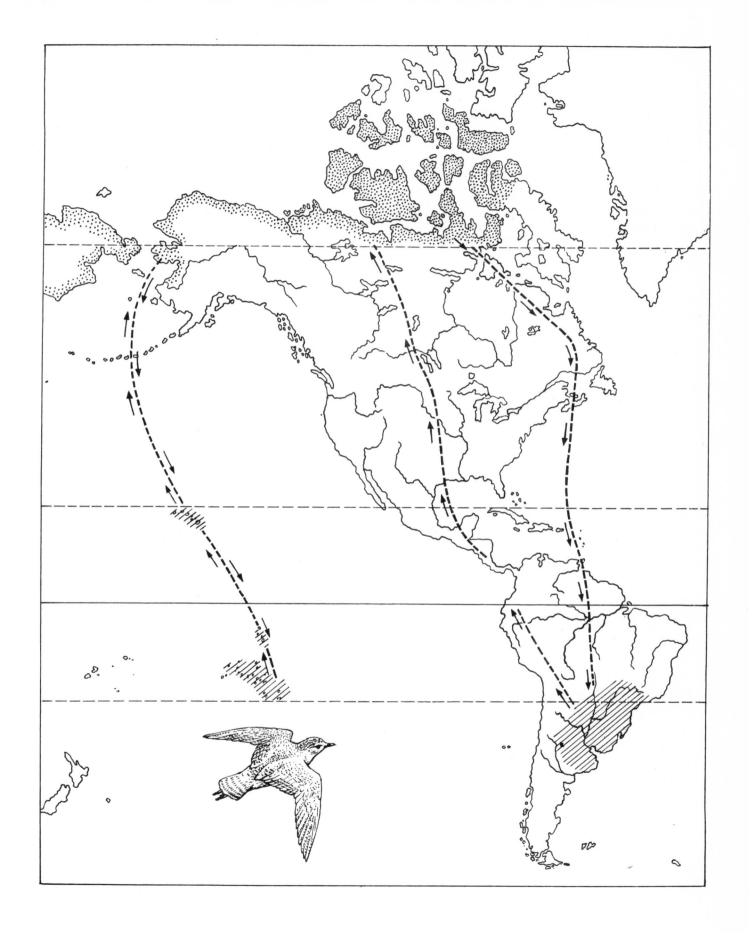

154

What remains of great consequence for all these relationships is the reminder of the above-mentioned author that the animals to be investigated are very complex organisms calling for research into their respective sense functions without the possibility of disregarding other mutually integrated functions. Finally, his point of view is, that the only findings that matter are those based on the situation in which navigation is really done, i.e. in free flight. "As for birds we still have no knowledge, or no satisfactory knowledge, concerning the way they perceive and measure magnetic fields, inertia forces, atmospheric and astronomical factors, time of day, time intervals, etc." Ostensibly any progress in this respect runs parallel to the rate at which the appropriate technical devices become available (radar implements, sub-miniature transmitters, vehicles, integrated electronic experimenting devices). All this calls for more and more means and has necessitated co-operation on the part of the respective specialists (physicists, engineers, statisticians, pilots).

In the ensuing pages we wish to remain true to our undertaking to give a popularized account of scientific phenomena. So we will stick soberly to the unequivocal empiric material hitherto assembled. Even so there is a lot to move us to admiration.

In the first place: orientation performances of migratory birds should be distinguished into direction orientation and destination bearing, and how vitally necessary a sense of direction can be is demonstrated by the wanderings of typical land birds across vast water surfaces. Any deviation, however minute, from the course would entail that the animals on their transoceanic passage would be doomed to perish. "Should the migration direction be a fortuitous one, then by far the greatest part of the birds would never reach their destination. This would result in substantial losses which no species could possibly survive." (F. Salomonsen, 1969) Once "on their way", the animals are thus constantly routed towards their destination. This sense of direction is passed on as a hereditary gift from one generation to another, and can, accordingly, be regarded as innate, impelling the bird to follow the traditional route throughout its migration passage.

As to factors governing these processes valuable information has been provided by transverse displacement experiments and experiments with migratory birds held in captivity. In the early fifties G. Kramer's experiments with migration-restless starlings in a circular cage in which the birds saw nothing but the sky showed that "species migrating by day possess the ability to determine innate or trained compass directions with the aid of the sun, and an endogenous clock indicating periods of day" (E. Gwinner, 1971). When a different incidence of the sun's rays was simulated by means of a mirror, the birds likewise took a different, predictable direction. When the sky was completely overcast they got confused as to their bearings. Thus the migration-active wanderers derived the knowledge of their flight direction from the angle of incidence of the sun's rays, and must have been capable of correctly appraising the sun's trajectory across the sky in the course of the day. This capacity results from the existence of an "interior" clock, a time mechanism which is synchronized with the succession of day and night due to earth rotation and "set" at the local time. This biological clock (which works on the principle of self-excited oscillations, and though it can be synchronized by exterior impulses is also capable of running on continuously even without such time indicators—cf. in K. Hoffmann, 1966) helps to compensate for changes occasioned by alterations in the sun's angle according to the time of day. Thus the migratory bird quite evidently calculates in the same way as a man who determines the cardinal points by the position of the sun and his watch-hand.

Birds heading for their winter quarters by night take, with a high degree of probability, their bearings from the constellations. Experiments conducted under an artificial sky in planetariums by the Sauers, an experimenting couple, with various species of European hedge-sparrows adduced that even there the normal migration routes were taken. Of course, a night sky when overcast played havoc with the birds' orientation; their performances in this respect then losing much of their precision.—It can also be assumed that orientation by the constellations is limited by a full moon as at least the fainter stars must fade in the bright moonshine.

147 Migration restlessness of a starling in a round cage. a—under an open sky, b—under an overcast sky, c and d—during the mirror test. During the latter light, measured on the normal incidence direction, is incident under an angle altered by 90 degrees. Each point symbolizes 10 seconds of directed "migration" in one of the 6 sectors of the cage. The thick arrows indicate the mean direction taken by the restless starling, the interrupted ones the path of the light.
After G. Kramer from E. Schüz.

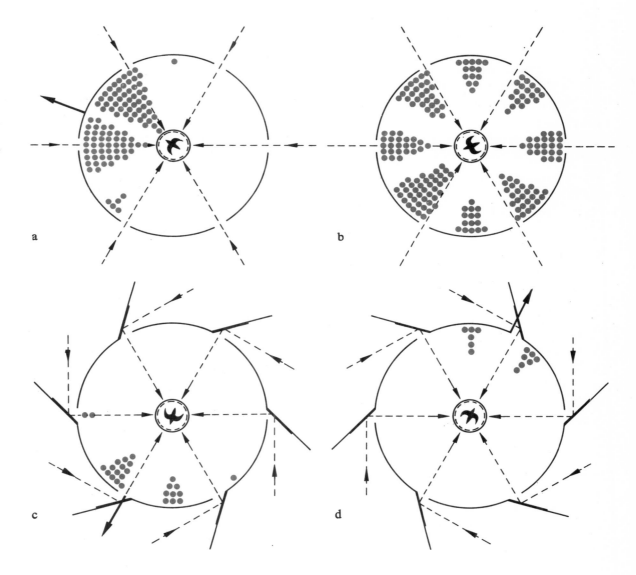

a

b

c

d

There have been growing indications recently—and this would be the third compass available to birds of passage—that our winged wanderers are capable of perceiving the earth's magnetic field and to make use of it as an aid to navigation. This might play an exceedingly important life-saving role when the sky is overcast and hence the constellations are obscured.

It can hardly be disputed after what has been said above that migratory birds take their bearings towards their destination. As for the How?—a question which is much more difficult to answer—quite a few pointers have been supplied by the many experiments carried out in more recent days. The alternative that offered was: do the birds reach their destination through groping from one landmark to another, or by making use of true navigation?

Accordingly, while carefully weighing the existing evidence we shall characterize the amazing orientation processes observed in birds as integrated effects of optical, time-measuring and central nervous elements whose exhaustive explanation is up to the future to supply.

148 Common cranes *(Grus grus)* on their passage to their hibernating quarters.

149 The autumnal migration of Nordic geese (here white-fronted geese) is one of the loveliest spectacles the Northern hemisphere can offer.

150 Migrating knots
(*Calidris canutus*) having a rest
on a lonely cliff.

151 The sight of such masses of
birds (knots at the west coast of
Denmark) lends credibility to the
showy phrase that the sun was
sometimes obscured by enormous
swarms of migrating birds.

152 Alpine marmots are long-hibernating animals; only every three or four weeks they interpose brief feeding breaks.

Regulation

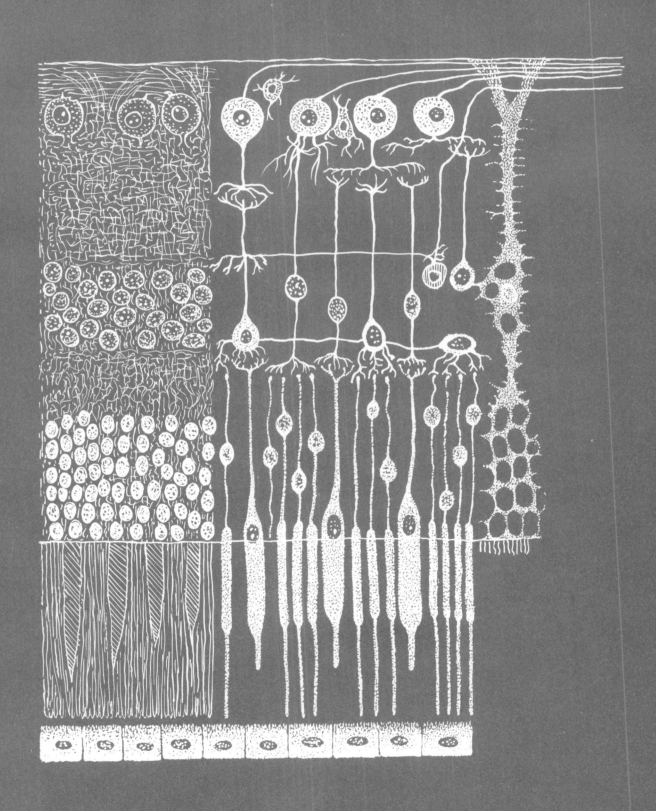

There ist hardly anyone who has not experienced the extremely vexing and unpleasant sensations called forth in the human body by fever. Also it is generally known that this fever is nothing but a few degrees' deviation upwards of body temperature from its normal value (37 °C). On the other hand, only few will know why and how it comes about that this normal value—except for slight variations due to the time of day—can be kept constant in a healthy human body.

Physicians and biologists have been aware for a long time now that in the highly specialized organism of warm-blooded animals with its sensitive cells body temperature, or more precisely blood temperature, is far from being the only quantity that must remain equal. The continued existence of life also requires that blood pressure, the CO_2 blood content, the hydrogen ion concentration, the blood sugar level and a number of other values may fluctuate only within a narrow range round the normal value. In the course of the phylogenetic development of the species regulating mechanisms developed working with great precision and efficiency by which—in complex interplay—all quantities essential to life are kept constant.

Even a reader who is little versed in technical matters will know that even in most every-day appliances and machines certain quantities must not fluctuate unduly if their satisfactory functioning is to be ensured. Such quantities are e. g. temperature in a refrigerator, in a motor-car engine, or an automatic washing machine; furthermore, tape speed of a cassette tape recorder, state of petrol in the float chamber of an automobile carburettor, and, which is less known from technology, frequency in a power network, speed of a steam turbine, gas pressure in a welding burner, temperature in a diffusion stove for producing microelectronic components, and many others.

Biological regulators have been in existence for millions of years now but it was but a few decades ago that their nature came to be recognized, and a set of concepts and a theory of regulation were developed in technology.—The first usable technical controller was the speed governor devised by James Watt (1736–1819) for his own steam engine. There has been a tempestuous upswing in control engineering since the beginning of the 20th century when entirely new possibilities were opened up by the development of the electron tube.

The conclusion drawn by Norbert Wiener (1894–1964), the ingenious American mathematician, from the many discussions he had had with medical men and natural scientists and from his own works was that regulation processes in biological and technical systems were based on the same principle and could be explained and expounded by one common theory. With his book entitled "Cybernetics" (1948) he founded a new scientific discipline—cybernetics—concerned not only with control and regulation but generally with ways of data processing and data transmission in animate beings and in machines. The term cybernetics is derived from the Greek word "kybernētēs" (steersman).

In the ensuing discourse the speed governor of James Watt is going to serve us as a vehicle for introducing some important basic concepts of control engineering by means of which biological regulators can also be understood and described.—The quantity that is to be kept constant is the rotational speed; this is called controlled condition. Disturbances of the rotational speed are caused by the fluctuating need of torque of the coupled machines, hence the term disturbance variable. Rotational speed, i.e. the number of revolutions, can be influenced by steam supply. This is determined by the lifting of the inlet valve referred to as control medium. Acting as a detecting element is the centrifugal-force tachometer which, with its system of levers for setting the valve—the final control element—constitutes the actual regulator. Controlled system—characterized by such concepts as controlled condition, disturbance variable, control medium, detecting element and final control element—together with the regulator make up a regulating circuit. A diagram of a regulating circuit is shown in Fig. 154.

Closed-loop automatic control takes place as follows: desired value for the controlled condition (rotational speed) is set fast (by means of an adjusting screw on the leverage between the centrifugal-force tachometer and the steam-inlet valve). The rotational speed actually existing at the moment, the actual value of the controlled condition, can be read from the position of the weights of the centrifugal-force tachometer. If the former lies below the desired value the steam-inlet valve is automatically opened farther by the connecting rod, in case it exceeds the value the steam supply is throttled in the same way. In this manner the regulator counteracts any deviation of the controlled condition from the desired value. This principle of counteracting (too high a rotational speed—throttling of steam input, too low a rotational speed—increase in steam input) is called "negative feedback". —It is characteristic of every regulator, and distinguishes regulation from control, which is a one-way process without any feedback. The lifting of the valve is determined by the position of the centrifugal-force tachometer—and the latter in turn by the rotational speed so that there exists an unequivocal interpolation between the valve lift and the rotational speed, a correlation by which regulation is made possible. Therefore, under normal working conditions, the rotational speed of steam engines can automatically be kept constant by this simple mechanical device working steadily and reacting to any change of the rotational speed.

In contrast to the continuous controllers which react to disturbances in the controlled condition adequately and without interruption, the discontinuous (intermittent) ones have only two positions for the final control element: ON (connected) and OFF (disconnected). Two-step controller is thus another name given to discontinuous controllers. These are to be found e.g. in thermostats used in refrigerators and automobile radiators, or in bimetallic thermostats of electric flat-irons.

Not infrequently the controlled condition of a device under differing working conditions is also expected to be able to reach various adjusted values—for instance, the temperature in an electric flat-iron when linen as well as synthetic clothes are to be ironed in succession. This is only possible by adjusting the desired value—the flat-iron has a rotary knob by which the setting of the bimetallic thermostat can be altered. The value affecting the desired value is called command variable (or reference input). It plays the decisive role in originating fever in the human and animal body as will be illustrated in the chapter on the organism of warm-blooded animals.

There are still two more terms of importance for control action: dead time and gain (or amplification). Dead time is the delay between the time a disturbance occurs and the time it is responded to by the controlled condition. Gain is a way of describing the

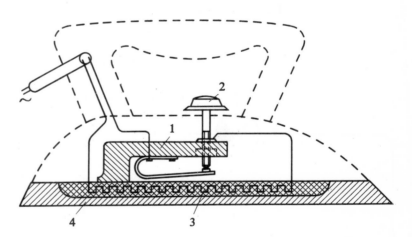

153 Electric flat-iron with two-point regulator. On the bimetallic regulator (1) the desired value of the ironing temperature can be adjusted by a turning knob (2) (e.g. for linen or man-made fibres). The coiled filament (3) heats the iron plate (4) indirectly. When the set temperature is reached the bimetallic strip deflects from contact on the preset knob axis, thus cutting off the current. Thereupon the flat-iron cools down, and the bimetallic contact closes the electric current for the heater spiral again, etc. After E. Samal.

164

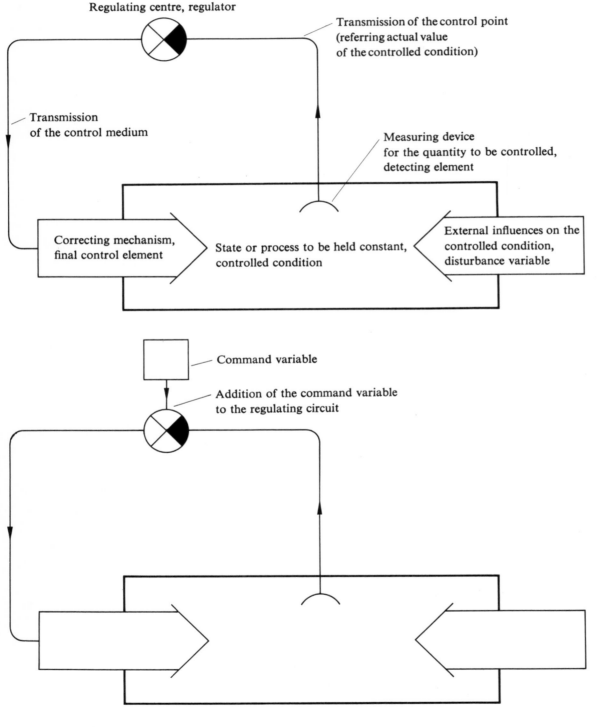

Regulating centre, regulator

Transmission of the control point
(referring actual value
of the controlled condition)

Transmission
of the control medium

Measuring device
for the quantity to be controlled,
detecting element

Correcting mechanism,
final control element

State or process to be held constant,
controlled condition

External influences on the
controlled condition,
disturbance variable

Command variable

Addition of the command variable
to the regulating circuit

154 Block scheme of a regulating circuit. After B. Hassenstein.

155 Block scheme of a regulating circuit which is affected by a command variable. Such a command variable is added to the desired value and thus amounts to an adjustment in the desired value. After B. Hassenstein.

156 Transition action of a regulating circuit with 1-second dead time for three different amplifications. The disturbance at t = 0 gives rise to a deviation in the controlled condition of —1. After 1 second dead time the regulator starts working. If the gain is 1, then the controlled condition is restored in oscillations with decreasing amplitude to the desired value. For gain V = 0.368 one reaches what is called aperiodic borderline case— the controlled condition approaches the desired value without overshot (asymptotically). On the other hand, if the gain exceeds 1—say 2—then the amplitude of the control oscillations does not decrease but increases: The regulator works in an unstable way and can destroy itself as a result of increased oscillations. After B. Hassenstein.

Thermostat and Regulating Circuit—the Organism of Warm-Blooded Animals

intensity of this response. E.g. gain amounts to 1 if after the inception of the regulation process set in motion by a disturbance the occurring control deviation (control deviation = the difference between the desired value and the actual value) gets compensated within one dead-time unit. If a shorter span is needed then gain is larger than 1, in the opposite case it is smaller than 1. If the gain is too high it may result in regulator oscillations which may be amplified into a regulator catastrophe—i.e. into its self-destruction. Regulator oscillations will be more thoroughly exemplified on the pupil of the human eye. When the gain is too small the desired value will not be reached until after a very long time.

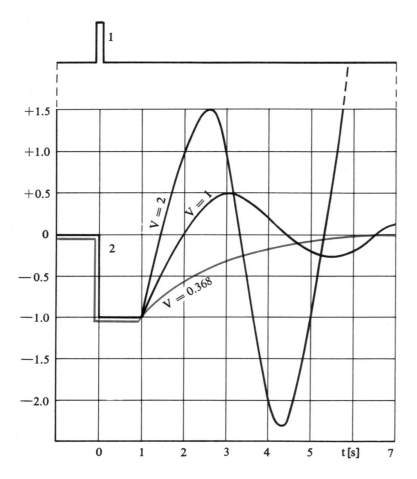

"Homoiothermic animals manage to keep their body temperature constant at an optimum level within a relatively wide range. This enables them to maintain the intensity of biological processes at its full strength when great differences and changes in temperature occur. It is only when in extreme ambient temperatures the control mechanisms fail that they become incapable of regulating their 'inner climate' so as to attain its optimum value. With the ambient heat at a minimum heat emission may exceed the production of animal heat to such an extent that body temperature drops below the optimum. As a result, the intensity of biological processes abates, and a cold paralysis or a 'benumbed state' occurs followed by death from cold unless the body is soon supplied with heat again. At very high ambient temperatures the amount of heat fed to the body may become so high that it can no longer be emitted by the physiological cooling mechanisms. Then body temperature exceeds its optimal level, which leads to disturbances in the organic functions followed by heat paralysis and death from heat." (K. Herter, 1956)

From the phylogenetic point of view the homoiothermic state is a comparatively recent development in the animal kingdom, and is known to exist in birds and mammals alone. Its essence consists in the fact that in alternating thermic situations the creation and emission of heat is invariably regulated by multiple control mechanisms in such a way that "as far as possible the genetically established index range of inner body temperatures is never abandoned." (G. Wittke, 1967)

This phenomenon is of a truly enormous biological advantage. It has enabled birds and mammals to settle our planet down to the least hospitable regions—impressive symbols of this being polar mews, penguins, polar bears, Arctix foxes, Antarctic wolves, snow leopards, Siberian tigers, mouflons of Alpine zones, etc. Moreover, it is interesting to note that not infrequently the homoiothermic species have to develop their capacity to regulate their body temperature "in a ready state" only after birth.

However, what actually is "temperature"? It is a variable of state, a measure for the heat content of solid bodies, liquids or gases. Yet temperature is not the same as heat, since heat is a form of energy stored in the motion of molecules or atoms (Brownian motion). To heat a body means to increase the kinetic energy of its smallest elements. When molecular motion equals zero or nearly so, then the material in question no longer contains any transferable heat—its temperature has reached the absolute zero point of 0 Kelvin (0 Kelvin = —273.15 °C). The speed of chemical reactions of which up to 2,000 can be taking place simultaneously in animal and human cells is strongly influenced by temperature; the former increases exponentially with the latter (Arrhenius equation). Though reaction velocity is very strongly affected by biological catalysts temperature none the less remains a significant factor.

Being convinced about the advantages of a temperature that is constant and not too low we will now trace the elements of the appropriate biological regulating circuit. The controlled condition is represented by blood temperature. The disturbance variables are easy to find; they consist in the cooling and overheating of the body caused by fluctuations in the ambient temperature, or by heavy work performed by the muscles. Final control elements are the blood vessels in the skin which can be contracted or extended, and thus either slow down or accelerate the transfer of heat from inside the body through the skin into the environment. Then there is the skeleton's muscular system which "produces" heat unaided by simultaneously stretching flexors and extensors (muscle tremor). Finally, another set of correcting elements are the sweat glands which, even at higher ambient temperatures, help to maintain body temperature at its normal value. Detecting elements are the receptors in the epidermis sensitive to heat and cold though of course their task in temperature control is merely one of forewarning when a rapid change in the ambient temperature occurs. Of greater importance are temperature-sensitive sensory cells which measure the actual blood temperature in the cerebellum itself. Nerve paths from the receptors to the CNS (afferent nerves) serve to transfer information by means of nerve action potentials. Commands to the correcting elements are also communicated in their turn, coded in nerve impulses,

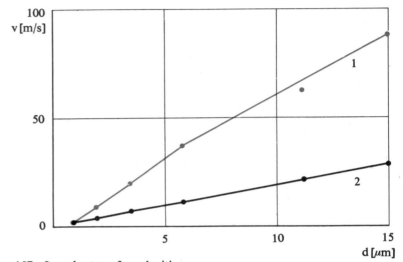

157 Impulse transfer velocities of nerve fibres in dependence on the fibre diameter d for warm-blooded animals (1) and poikilotherms (2). After E. Schubert.

over nerve paths (efferent nerves). The normal value of body temperature is pre-coded as the desired value in the regulator differently for each species of warm-blooded animals. Among mammals this lies between 32 °C (spiny anteater) and 40 °C (goat). Birds generally possess a higher body temperature than mammals, its increase being noted with the decreasing size of the body: kiwi 37.8 °C, domestic fowl 41.7 °C, and robin 44.6 °C.

If the temperature detecting elements report a deviation from the desired value, the regulator gives an order to the correcting elements to counteract this deviation. Too high a temperature is responded to by "cooling", one that is too low by "warming up". Thus what one can see being repeatedly implemented here is the principle of negative feedback which is indispensable to any kind of regulation. A change in blood temperature in the "body core" (the head and the digestive tract) hardly ever occurs at all.

The entire regulating circuit is so devised as to ensure a high reliability, many elements (e.g. detecting elements) being provided in more than one set. In addition, the thermo-regulator interacts with other regulating circuits (that for blood pressure, etc.); of course, the task of keeping body temperature constant is given priority.

Even from these few remarks on biological regulators it may have become clear that the latter though working on the same principle and lending themselves to the same description as the technical ones are characterized by a substantially more complex structure and are fully integrated in the living organism.

But to return to the way the biological thermostat works. What physical and chemical processes are available to the warm-blooded organism for heat exchange with environment and for the heat production? Let us first turn our attention to heat exchange. There is heat radiation we have already come across in describing the ultra-thermometer of the pit vipers. The phenomenon is one of extreme importance for heat economy: the body radiates heat energy in the form of electromagnetic waves with a wave-length of about 10 μm. On the other hand, it is capable of absorbing heat in this way—a pleasant experience known to everyone from sunbathing. Further, one should mention heat conduction which is of importance firstly for heat transmission from the interior of the body to its surface—here let us only recall the isolating layer of fat in whales—and secondly has a role to play in heat transmission in stationary air. Conduction of heat is, of course, possible only in terms of a temperature drop from higher-temperature ranges to those with a lower one. Finally, there is convection, during which heated air is carried off by the flow. This is encountered e.g. in breathing. Convection is also responsible for leading off heat from the body surface though this is strongly impeded by hair or feather covering. Of first-rate importance for heat radiation is, however, vaporization of liquid—either in the form of perspiration from the body surface, or as saliva from the open mouth in panting (the dog, who has no sweat glands, has to rely on this). This automatic process as the only one among those mentioned above makes it possible to radiate large amounts of superfluous heat even when ambient temperature lies considerably above body temperature (30 to 50 degrees) provided relative air humidity is not too high. Over 500 kilogramme calories, which is an amount of heat which can bring about 5 litres of water to boiling from zero, are taken away from the body in secreting 1 litre of sweat. About 5 litres of this salty liquid are secreted by a man's sweat glands on a single day!

Heat transmission inside the body is predominantly taken care of by blood which flows even through the remotest parts of the body.

Thus homoiothermic animals dispose of a rich inventory of physical and chemical regulating devices to defend themselves against "improper" cold and heat. The working of both is very finely attuned to each other. It can be observed that chemical regulation sets in when the physical means become overstrained. Thus they constitute autonomous components of thermo-regulatory reactions in a warm-blooded animal. These are to be distinguished from behavioural regulations to which one can ascribe such phenomena as seeking the shade, wallowing when it is very hot, restriction of movement, crowding or flocking together and preference for spots protected from the wind during a severe cold, but also long-distance migrations.

158 The speed controller designed by James Watt for his first steam engine dating from the year 1788 (a replica). The two spherical weights (balls) of the centrifugal-force tachometer are to be seen in the middle of the picture. The rods extending to the left of the axis of rotation of the centrifugal-force tachometer actuate the steam inlet valve of the machine in dependence on the balls.

"Such ways of behaviour are on the one hand suited to relieve the 'interior' control mechanisms and to economize their operation while, on the other hand, once their correcting ranges have been exceeded, possibly to ensure the organism its last refuge from the threat of freezing or overheating." (cf. in G. Wittke, 1967)

As for the physical preventive means or measures, these are, roughly speaking, oriented in a peripheral way. Thus they manipulate the shell of the body, its surface, in order to prevent a drop in inner body temperature by altering the heat-conduction capacity of the former. This set of instruments includes alterations in blood circulation of the skin, sweating, panting, thermal isolation by seasonally renewed thick layers of feathers, furs or blubber. It is so effective that e.g. Arctic foxes and polar bears need to enhance their basic "inner" metabolism only when the temperature in the Arctic surroundings has dropped to —40 °C or —50 °C.

Chemical protection means against cold amount all in all to "boosting up" the inside of the body. Cell metabolism gets increased—thus producing more heat—and the preceding secretion of metabolism promoting hormones into the circulation tends to further this process to a substantial degree. "In particular the medulla and the cortex of the suprarenal gland as well as the thyroid gland become activated." (Raths and Biewald) Muscle tension increases; this may find its expression in "chattering teeth" and "shivering with cold".

There remains the question of the place where all these thermoregulatory processes are integrated in the organism of warm-blooded birds and mammals. The anterior *regio hypothalamica*, the forward bottom of the interbrain (diencephalon) is generally regarded as being the place, and understandably zoologists and medical men have long been at pains to explain to the minutest detail the regulating function of this nerve structure. We may rest satisfied with the finding that the reports on actual temperatures in both the shell and core reach not only the cerebral cortex but also the interbrain (diencephalon) and in its hypothalamus we are confronted with a controller which constantly compares the existing actual value with the thermal desired value at which it is set in the hereditary process and, if required, intervenes in a correcting way.

Nevertheless, in spite of the precise and demanding regulation it does occasionally occur that body temperature deviates from its normal value. This can happen when the disturbance variables (ambient temperature, muscle labour) take on such values that compensation by means of correcting elements is no longer possible. These phenomena are known as hypothermia (undercooling) and hyperthermia (overheating). If the disturbance is too powerful the regulator has no effect.

Entirely different circumstances obtain when fever occurs. The regulating circuit is then completely intact, only the desired value in the regulator is affected by a command variable which adjusts it to a higher temperature value. This new desired value then also becomes adjusted with great precision. At the same time the regulator "endeavours" to bring body temperature as soon as possible to the new value. The result of this endeavour known to everybody are heat-producing shivers at the outbreak of a febrile illness, and the powerful sweats at its end. The cause of deviations from the desired value are bacterial toxins, or outright injuries to the regulation centre. Therefore fever can be removed only by medicaments which directly affect the CNS and eliminate the command variable. Compresses can act only as a temporarily easing disturbance variable.

159 The thick fur helps the polar bear living in Arctic latitudes to avoid a perilous drop in body temperature.

160 Steadily evaporating sweat
draws superfluous heat from the
body of a foundryman.

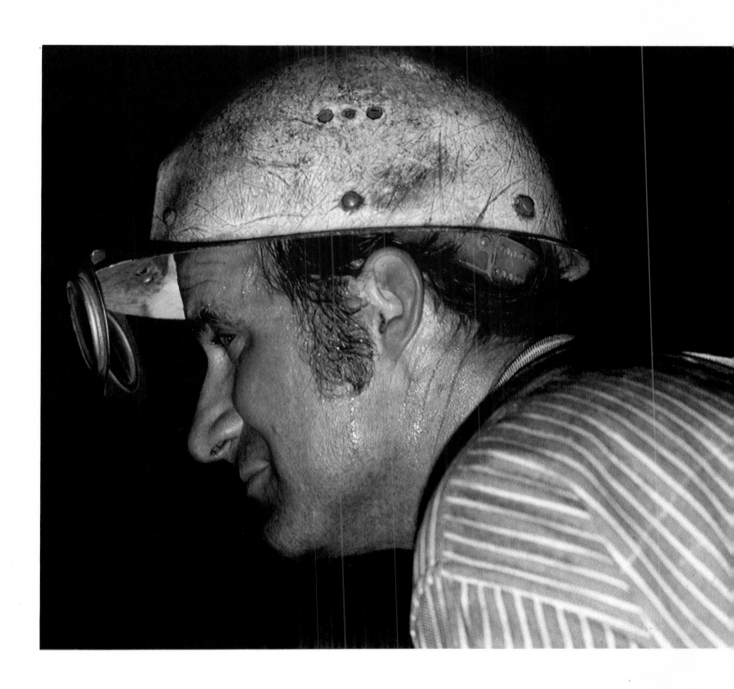

161 Isolating fur also protects the organism of warm-blooded animals such as the young harp or saddleback seal *(Pagophilus groenlandica)* from undercooling in its arctic environment.

162 The Fiji-leguan—a bright-coloured representative of the poikilothermic lizards.

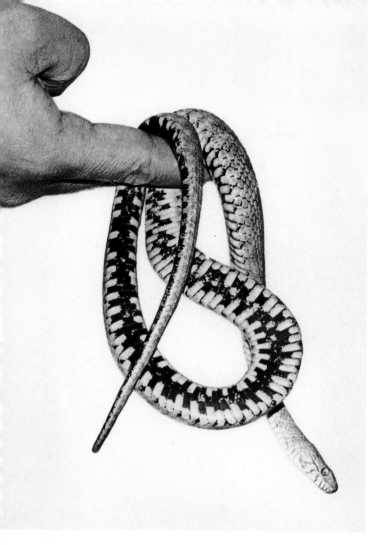

163 Scorched by the sun is this landscape in Somaliland.

164 Gardener's work in late autumn or early spring sometimes brings to light a cold-rigid snake. This dice snake needed some time to "thaw".

165 Drought in Sahel. No rain for years and the heat have claimed their victims among plants and animals.

How Do Poikilothermic Animals Protect Themselves against Too Great Heat and Cold?

Poikilothermic animals are animals with a variable body temperature, i.e. living creatures which under compulsion share the fluctuation of temperature in their environment. Homoiothermic animals, i.e. those with uniform body temperature, are better provided for in this respect; they are capable of regulating heat absorption and emission in such a way that, in spite of variations in ambient temperature, their own body temperature remains essentially constant.

Those referred to as poikilotherms include insects, fishes, batrachians, reptiles, while homoiotherms are birds, mammals, the latter including man.

A plastic illustration of what is meant by poikilothermia is given to anyone who has ever tried to catch with his hands lizards that had stayed on a sunny spot for any length of time. Outright diaphoretic tactical manoeuvres have to be undertaken in order to catch the warmed up, and hence extremely agile, little reptiles before they whizz and disappear in their hiding places. One's chance of success is greater when they have only just left their dens in the morning and are still a little "cool".

Or another example: "A moth or cockchafer that is taken off a leaf after a cool night hardly moves at all. Its limbs are stiff with cold, its wings are out of use. In spite of this, it is just moths and cockchafers that are active by night. To be able to fly in cold weather at all they raise their body temperature before taking off. They buzz with the wings and increase oxygen supply by pumping body movements. The powerful muscular activity produces 'animal heat' and physical frictional heat. With ambient temperature at 11 °C a butterfly gets heated from 11 °C to 33 °C in six minutes.—Poikilothermic animals with large muscles find it easier to produce considerable amounts of metabolic heat. Consequently, giant boas and giant pythons have a higher temperature than their environment and many of them make use of it to hatch their eggs with. The tunny even reaches temperatures exceeding that of the water by 10 °C to 14 °C. Likewise honeybees as a whole swarm possess rather a great amount of muscles, hence they produce great amounts of heat, and attain a hive temperature of 35 °C which is most favourable for the development of the larvae."

The lesson derived from the above words written by the physiologists P. Raths and G.-A. Biewald is that many poikilothermic animals are not quite so powerless when faced with variations in environmental temperature. As a means of counteracting too low and too high temperatures they use special ways of behaviour, physiological regulations and adjustment on a biochemical basis.

Life without heat would be impossible since chemical processes in animate beings take place under absorption or emission of heat. Most metabolic processes give rise to heat in the animal body which the poikilothermic animals—representing by far the greatest majority of animals—rapidly diffuse into their environment. This makes it easy to understand that in examining most varied vital processes such as heart beat and breathing frequency, frequency of the cerebral currents, conduction velocity in nerves, oxygen consumption etc. their dependence on temperature becomes only too clear. All vital phenomena based on metabolism give evidence of this nexus, and the reason is simply to be sought in the fact that reaction velocity in chemical conversions grows with rising temperature (Arrhenius equation). Once the metabolic processes are set in motion, they become more intensive and rapid with rising

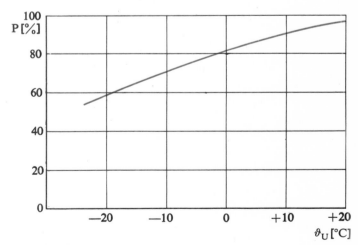

167   Effect of ambient temperature on performance P of a lead storage battery.

temperature up to an optimum which represents a quite particularly favourable thermal "niche" in its habitat for each individual animal.

"However, the majority of the poikilotherms absorb the heat necessary for the vital processes from outside, i.e. through radiation and conduction from the medium, the substrate and the base. Temperature conditions in natural biotopes are as a rule so fluctuating in locality and time that it is but a rare occurrence that an animal finds itself in each spot of the biotope and for a long time in its optimal temperature ... Generally," K. Herter goes on, "animals are compelled to use active motion to 'find out' such places whose temperatures approach their own original temperatures as closely as possible. This they are capable of doing with the aid of the thermotaxis which leads them to places with a certain condition—i.e. to their P.F.-area (preferred-temperature area—authors' note), or keeps them there. It is to be assumed that the P.F.-area is a temperature zone in which, as a result of radiation or conduction so much heat is absorbed or lost by the body that it takes on original temperature adequate to its existing physiological state, or keeps it up for a time. Here it needs to produce less heat than at lower temperatures and thus saves energy which at optimum-exceeding temperatures would have to be expended for regulating its own temperature (e.g. by perspiration). Therefore in the preference-temperature area the intensity of metabolism will be smaller than at lower temperatures or higher ones."

Animal life exists only within a relatively limited temperature range. It would imply death if the humours of the body should solidify—for, indeed, these are nothing more than water solutions of various substances which can be physiologically active only when preserved in their fluid state. Nevertheless, in undercooling the standstill to which vital organ actions come occurs later with poikilothermic animals than with homoiothermic ones. Temperatures down to zero Centigrade are hardly dangerous as a rule. For instance, insects, snails, frogs, lizards outlive the winter in torpor at temperatures close round the freezing point. As a general rule, poikilothermic animals counteract the threat to their lives by increasing their resistance to low and high temperatures. Polar fish, which live in water that is 5 °C "warm" in summer and in which the freezing point of the blood plasma is 0.8 °C lower, shift this freezing point to —1.5 °C in winter as soon as the temperature of the surrounding medium drops to —1.8 °C. In this way the danger of freezing is reduced. Indeed, should the water in the cells solidify to ice crystals this would result in the destruction of the cell structure, and the animals would inevitably perish. Cells are preserved by the cell sap becoming thickened, hence by dehydration of the tissues. This phenomenon is found to be generally current among insects as well. The eatable mussel heightens the frost resistance of its body tissue by "adding" some kinds of sugar or glycerine to its water. There are insects which are capable of bringing the freezing point of the humours of their body down to —17 °C in this way.

Naturally, this kind of adaptation is also made in the direction of increased power of resistance against heat, in which either the body as a whole but also its individual organs can participate. When a frog begins to feel too warm on its white water-lily leaf, it will either jump into the cooling water, or at least move on a bit farther on its base where the belly and the extremities are once again more comfortably tempered. The animal is sure to have looked in the morning for the warming rays of the rising sun in order to make itself internally ready "for tours". At 15 °C its heartbeat was already three times faster than at 5 °C, and at 25 °C again three times as fast as at 15 °C. Of course, should the temperature exceed 30 °C, then the whole animal's habituation and the thermal adaptability of the cell enzymes (agents) would no longer cope with the situation. If our little frog had not meanwhile been "sensible" enough to move into the shade—which it is, however, certain to have done—metabolism and the functioning of a number of organs would get blocked up by too high a temperature. At 44 °C the frog heart would die of overheating: the cell protein would begin to coagulate—together with other vital organs—and the enzymes would thus suffer destruction.

Anyone given to deep and sound sleeping is popularly described in German as sleeping "like a marmot" (the English equivalent being "like a log or like a top"). And many a man may like to associate this saying with the idea of winter sleep, as it is just common knowledge that in our latitudes the hedgehog, the marmot, the hamster and the dormouse live through the cold season of the year sleeping under the surface of the earth or in some overground remote corner. In the now classic words of M. Eisentraut (1956) what nature demonstrates here on some warm-blooded animals is an experiment deserving notice and admiration: an animal which in its active state is extraordinarily lively and agile is, once exterior conditions unfavourable, or even menacing, for the preservation of life appear, submerged into a profound lethargy in which hardly any sign of life can be perceived, only to re-awaken from this condition of seeming death to a new life.

It will be good to make a clear distinction between this rigid state and similar phenomena, since in the animal kingdom several varieties of inactivity and rest are known. According to Eisentraut, here the body is in a state different from normal sleep that finds its expression in more or less complete lethargy. "Depending on the profundity of the animal's state the outwardly noticeable lower breath frequency and pulse rate are decreased, the whole metabolism gets reduced and nervous activity is largely eliminated. This functional deceleration is most closely linked with a drop in body temperature. Body-heat regulating mechanisms are reversed to a certain degree. The body is in a condition of rest and in a state of rigidity, the animal's capacity of motion being arrested. Food reserves stored in the body serve to maintain the 'vita minima'. Waking from the lethargy which normally occurs in many species periodically within the hibernation time, can be actuated by exterior or interior stimuli and takes rather a long time to accomplish. Rapid combustion processes starting in the body subsequently bring about a rise in body temperature, all vital functions becoming gradually restored. Only after reaching the lower threshold of waking temperature the animal is once again capable of making an unimpeded use of its organs of motion, and finally returns to the active way of life."

This precise characterization makes us easily understand that the winter sleep phenomenon is tied up with certain physiological prerequisites. If these are fulfilled only approximately or partially, we have no right to speak of hibernating animals. Of course, it may be conceded that "there can be a variety of transitions and intermediate degrees ranging from winter rest, occasional short-term semi-concious states and drowsiness with a light drop in temperature, a transitory coldness lethargy up to real winter sleep with an outright long-lasting lethargy."

168   A dormouse during awakening from its winter sleep.
Left: The animal in deep sleep.
Centre: The sleeping condition relaxes.
Right: Short before awakening, the eyes are still closed.
After M. Eisentraut.

(cf. in M. Eisentraut) Even in "true" hibernating animals their behaviour gradually varies: "long" and "short" sleepers are known to exist. Marmots and some dormice are long sleepers—their lethargy lasts for weeks, indeed for months—, while for instance, hamsters wake up repeatedly after a few days and seek their subterranean larder to consume something of their piled-up provisions.

More recent researches have shown that hibernating animals occur in a number of orders of mammals—in monotremes (Monotremata), marsupials (Marsupialia), insect eaters (Insectivora), bats (Chiroptera), rodents (Rodentia) and primates (Primates, suborder Prosimii).

In order to illustrate the substance of the matter and to demonstrate the regulatory phenomena on an example, let us take a marmot and a hamster as objects of our experiment. To begin with, however, a few words on the biological significance of our phenomenon. It is in the first place the scarcity of nutrition setting in with the coming of winter, and not so much the appearance of cold, that causes the temporary renunciation of the active way of life. The drop in body temperature at the same time entails a reduction in metabolism; thus the cold, lethargic hibernating animal needs only a very little amount of nourishment. The stored-up body fat, or the supplies gradually carried into the winter burrow, are sufficient to keep the vital functions going—even though considerably reduced. Therefore P. Raths and G.-A. Biewald were entirely consequential in describing winter sleep in general as a luxury which makes life in bad times more comfortable. They point out that it was able to develop where biochemical preconditions made it possible. Hence the problem becomes one of heat regulation and as such calls for a more detailed explanation at a later stage.

In the first place, however,—to illustrate the matter more clearly—we will keep a marmot and a hamster under observation. What do they do as soon as winter approaches? The winter lethargy does not enter their lives without some intervening means. With the coming of autumn outside temperature drops step by step, and this is what, to a certain extent, marks the time of preparation of both the rodents for the long period of rest. The animals "lose their vivacity, become sleepy and lazy ... If at the outset of hibernation body temperature sinks below the lower limit of waking temperature and below the activity threshold, then a light lethargy sets in, which, the more the body heat decreases becomes ever deeper until, finally, when minimum temperature is approached the deepest state of winter sleep is reached. During awakening as a result of re-warming, the above way will be traversed in the opposite direction until finally, on reaching the lower limit of waking temperature, the animal resumes active life again." (see M. Eisentraut)

Marmots generally use special winter nests which are often an extension of the burrows inhabited in the summer months. At the end of a longer corridor there is a spacious round-shaped hollow, "cosily" upholstered with dry plant materials. It often lies several metres under the surface of the earth. Alpine marmots bring considerable quantities of hay into their winter burrows. Hamsters, too, enlarge their summer burrows when autumn passes and build little spaces for winter sleep and supplies. In their summer sleeping nests only the ground of the hollow is covered with upholstering; winter sleeping rooms are completely filled with it. When hamsters were dug out, they were found to be thickly enveloped by it on all sides. Marmots prepare food reserves by mainly storing up body fat, thus practising what might be called "physiological reserve economy". Hamsters practise an "ecological" one, furnishing their winter quarters with food supplies. This becomes understandable just because one knows that hibernating hamsters do not "sleep through" 3 to 4 weeks the way marmots do, the former having substantially shorter periods of sleep than the latter. When the hamster awakes about every fifth day it must take some food, for it has just strongly heated up its body and thereby consumed its inner reserves. Then it falls asleep again. For developing fats in the lie-abeds (cf. in M. Eisentraut) inner physiological readiness plays the decisive role; it has little to do with the rapidly increased quantities of food following the approach of autumn. For instance, dormice held in captivity regularly become stout without being given more food. Naturally the fat cushions raise body weight considerably; in a marmot 900 grammes of fat were established in autumn at an average weight of 3,000 grammes. When the animals re-appear in front of their burrows in spring, the reserves are more or less used up, their emaciated appearance being unmistakable evidence of this.

The winter state of lethargy is brought about by a combination of interior (endogenous) and exterior (exogenous) factors. Only their interplay can give rise to this state; these are then the mechanisms of hibernation. As exterior factors one has to name being shut off from light, cut off from fresh air, deprivation of nourishment, the qualitative change in food towards the end of summer and the autumnal drop in ambient temperature. The last named factor is decisive for initiating hibernation; however, it is not its only cause. The above-mentioned accessory circumstances and further endogenous factors are essential for making the animal ready for hibernation.

The set of effects is completed by reserves of fat, changes in the endocrine glands, in hormone economy and the central nervous control. The most far-reaching effects that dispose marmots, hamsters, hedgehogs, dormice etc. for winter lethargy are functions of the endocrine glands.

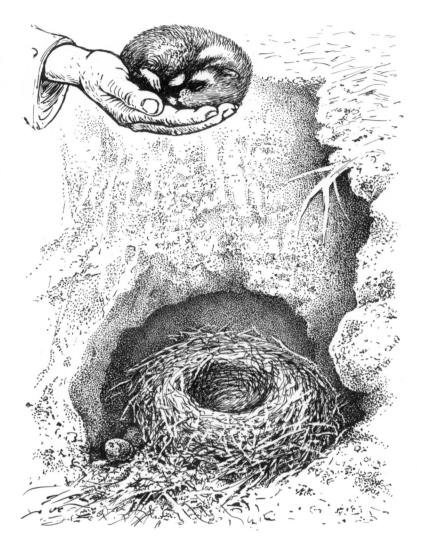

169 The open winter burrow of a hamster—the sleeping animal was taken out of the nest. After M. Eisentraut.

170 Ground-plan of a winter burrow of a hamster. Coloured: Partly obstructed summer burrow. After M. Eisentraut.

182

171 Rise in body temperature ϑ_K with time for a golden hamster who awakes from its winter sleep. After M. Eisentraut.

172 Body temperature ϑ_K over ambient temperature ϑ_U (schematic) for a poikilothermic non-hibernating animal (1), a hibernating animal (2), and for a warm-blooded non-hibernating animal (3). While the body temperature of a poikilothermic non-hibernating animal drops synchronously with ambient temperature until the humours of the body freeze and exitus (death) follows, body temperature of hibernating animals remains constant until the lower limit of waking temperature is reached. Then the animal slowly falls into its winter sleep and now its body temperature also

proceeds synchronously with ambient temperature, only so long of course until a life-endangering temperature value is reached. Then the animal wakes up and its temperature rises again. On the contrary, warm-blooded non-hibernating animals must keep their temperature constant under all circumstances. If it drops only a few degrees below the desired value, death follows. After M. Eisentraut.

173 Loss of weight with time incurred by dwarf ground squirrels.
1 when hibernating, 2 going hungry while awake. The drawing drastically indicates the advantages of winter sleep during the cold and when food is scarce. After M. Eisentraut.

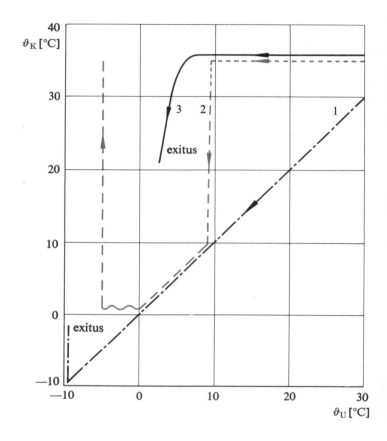

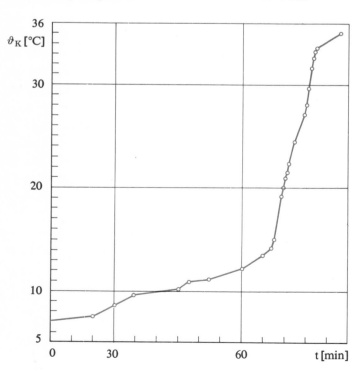

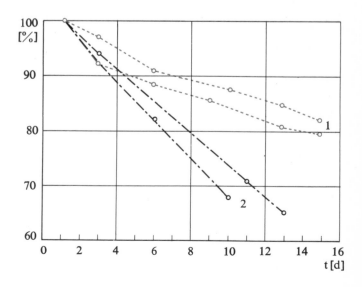

The Pupil—a Model of a Cybernetic Regulator

Above winter sleep had been described as primarily a question of heat regulation. This is subject to regulation by the interbrain (diencephalon) or the hypothalamus, which acts as a thermostat in many lethargic hibernating animals. The animals' daily rhythm, their waking state and sleep, gradually ceases and—as soon as the outside temperature drops to the corresponding degree—their daily sleeping periods pass smoothly into the lethargy of winter sleep: i.e. a thermostat and regulating centre in one. Its closest neighbour is the pituitary gland, the commanding post for endocrine secretion. Thus it is seen that nervous and hormonal impulses not only act together in fine unity but are also closely interconnected externally.

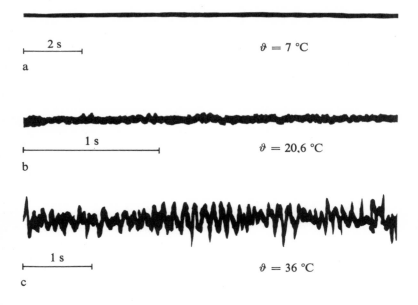

2 s $\vartheta = 7\ ^\circ\text{C}$

a

1 s $\vartheta = 20.6\ ^\circ\text{C}$

b

1 s $\vartheta = 36\ ^\circ\text{C}$

c

174 Electroencephalogram of *Citellus citellus* (common souslik) in the course of awakening. After M. Eisentraut.

The character of the face is determined by the size, shape and colour of the visible parts of the eye—this is equally true, with few limitations, of man as of other vertebrates. What is more, for man the sense of vision inherent in the eye is the most important of all senses: we are visual creatures par excellence. But who, in contemplating beautiful, expressive eyes, would hit upon the idea of making comparisons with technical structures, or even of searching for physical and cybernetic explanations? And yet this very eye of a vertebrate is a rewarding object for the physicist as well as for the control engineer. The pupil, its natural diaphragm device, is a model of a cybernetic regulating circuit upon which analogies between natural and artificial control systems are particularly easy to demonstrate. However, prior to dealing with this interesting biological regulator we wish to call to mind some important elements and properties of the human eye.

Held by six muscles, the eyeball is suspended in the eye socket revolving in all directions like a socket joint. The sectional drawing illustrates its inner structure. The entire optic system—pupil, lens, retina—is equal to one of a photographic camera. Just as in the latter a diminished reversed image of the perceived object is produced and projected on the light-sensitive retina—just as there on a light-sensitive film. Light quanta (photons) excite receptors which in turn communicate the optic information coded in the form of nerve action potentials by the optic nerve to the visual centre in the brain where the picture is again put together and evaluated. The retina is filled with a thick succession of rods for the light-dark vision and the less sensitive cones for colour vision. In the retina pit (fovea), the spot of the keenest sight, the distance between two neighbouring cones is only 4 to 5 µm.—Sharpness of sight is usually characterized as the capability of distinguishing two point-shaped small things still as separate objects. The human eye is still capable of separating two rays of light which enter through the eye junction under the angle of only one minute of arc (R. Hooke, 1764). Such an angle is formed by two rays of light which depart from two points 0.2 millimetre distant from each other at a distance of 1 metre from the eye. The limit of optical resolving power during a given wave-length is determined by the diameter of the pupil and the inner focal distance of the crystalline lens. As a result of the wave nature of light diffraction disks are formed on the retina by the optical system whose diameters depend on these two quantities. If the diffraction disks of two light rays overlap they can no longer be perceived as separate. In actual fact diffraction disks at a pupil distance of 4 millimetres and an inner focal distance of the lens of 1.7 centimetres have a diameter of about 4 to 5 μm. A smaller distance of the cones would amount to wasting as it could not lend the eye any higher resolving power. The retina and the optical system of the eye seem to be optimally attuned to each other.

On the other hand, lack of image definition grows with the third power of the aperture ratio (aperture ratio = pupil diameter divided by focal length of lens.) With good lighting the pupil diameter is small, and hence the pictures are sharper, i.e. more sharply defined. Therefore sufficient illumination is a basic precondition for filigree-work. The influence of aperture diameter on image definition is currently known to us from the camera: at small aperture diameters (high aperture number, e.g. 11 or 16) a substantially greater depth of focus is obtained than at large ones.

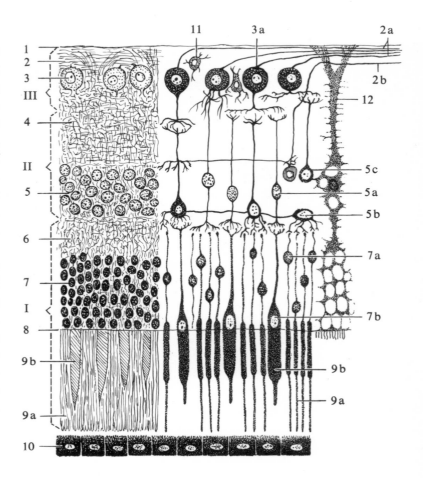

175 Section through the human eye.
1 Ciliary muscle with crystal lens,
2 Iris with pupil,
3 Sclera,
4 Retina pit (yellow spot),
5 Blind spot (entrance of the optic nerve),
6 Retina,
7 Vitreous humour,
8 Cornea,
9 Anterior chamber.

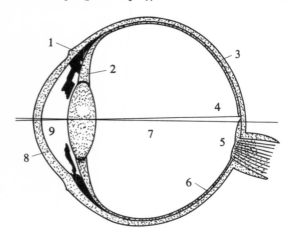

176 Vertical section through the vertebrate retina, its elements and their connection:
I Neuroepithelium layer,
II Ganglion retinae (layer of bipolar ganglion cells),
III Ganglion nervi optici (layer of large ganglion cells).
1 Membrana limitans interna,
2 Optic nerve layer,
2 a Centripetal fibres (neurites from 3 a),
2 b Centrifugal fibres,
3 Ganglion cells,
3 a Large ganglion cells (neurons of the optic nerves),
4 Inner plexiform layer,
5 Inner nuclear layer,
5 a Bibolar cell,
5 b Horizontal cell,
5 c Amacrine cells,
6 Outer plexiform layer,
7 Outer nuclear layer (layer of optic cells)
7 a Rod-optic cells
7 b Cone-optic cells
8 Membrana limitans externa,
9 Layer of rods and cones,
9 a Rods,
9 b Cones,
10 Pigment epithelium,
11 Glia cell,
12 Supporting radial fibres of Müller.

After R. Hesse from H. Giersberg and P. Rietschel.

177 When the strength of the illumination in animal environment changes it is the pupil that functions as the first antidazzling device: octopus, *Strix uralensis*, night-ape, *Profelis aurata*.

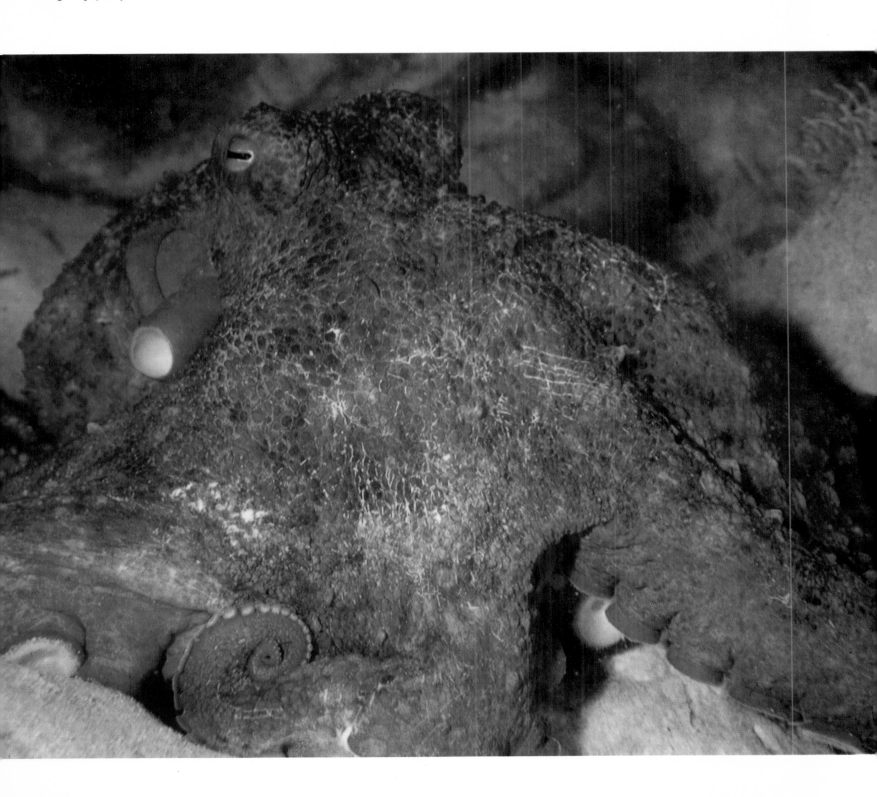

The adaptation range of the human eye can hardly be imagined.

The ratio of illumination intensities at which this important sense-organ still works satisfactorily is about 1:100 thousand million (1:10^{11})! Such extreme differences in lighting exist between the night—when the eye still offers us a weak black-and-white image (scotopic vision mediated by rods)—and a sunny day. Simple reasoning shows at once that the pupil is not in a position to keep the luminous flux (luminous flux = illumination intensity times pupil surface) into the eye within such a wide range even approximately constant. Its diameter can actually have only values between a minimum of 2 millimetres and a maximum of 8 millimetres. Since the luminous flux of the pupil surface is proportional it can be varied in the proportion of 1:16 at the most. Thus during changes in illumination the pupil is essentially an effective first anti-glare device.

Far more is achieved by the retina receptors. These are adaptable within a very wide range, of course for this they require more time than the pupil does. Many an amateur photographer may already have wished that the sensitivity of the photographic emulsion might be just as variable as the retina of the human eye when he was unable to take a unique motif because the exposure time and the aperture were not sufficient for the inserted film in the prevailing light conditions. Accordingly, adaptation characteristic of sense-organs constitutes a substantial advantage of natural receptors.

But to return to the pupil regulating circuit. What will be of particular interest to us are instabilities of the control system which can be very problematic in technical regulators and can be produced with relative ease with the pupil as uncontrolled oscillations in the pupil width. We know from the introductory chapter that the stability of a regulating circuit is influenced by dead time and gain. Both have to be exactly attuned to each other. If gain is artificially raised in a stably working regulator with predetermined dead time the oscillations occur in the controlled condition (see Fig. 156 in the introductory chapter). And such oscillations can be imposed on the pupil by a simple trick by means of which the gain of the pupil's regulating circuit is raised from outside. This effect was described by B. Hassenstein (1967). Before explaining his experiment let us first

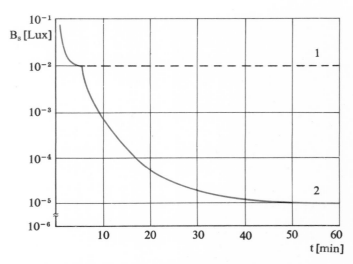

178   Decrease in threshold intensity (under threshold intensity one has to understand that intensity of illumination [B$_s$] at which the eye just reacts) of the retina of the human eye with time during dark adaptation. After E. Schubert.

179   Pupil diameter [d$_P$] of the human eye in dependence on intensity of illumination [B$_s$] in the main range of adjustment. After values given by E. Schober.

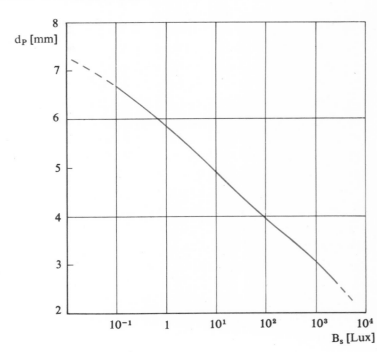

190

characterize the pupil's regulating circuit in a little more detail. The controlled condition is the illumination on the retina, which is determined by the luminous flux into the eye. Disturbance variables are the changes in lighting conditions. As detecting elements function the light-sensitive receptors of the retina on whose surface every spot is suitable for this purpose. If those deviations from the (variable) desired value are communicated to the regulator in the mid-brain (mesencephalon), an order is given to the final control element—the iris muscle system—to counteract the deviation from the rule. The pupil is contracted, or extended.

Now how can it be done to raise the gain of the pupil's regulating circuit from outside to such a degree that self-excited oscillations of the pupil's width may occur?—All one needs for this is a uniformly lit surface which can be observed by the patient and a glass disk with a circular black disk stuck on it, whose diameter is a little smaller than the diameter of the pupil in the prevailing illumination. If the patient is now asked to cover one eye and to put the glass with the paper disk exactly so in front of the free eye that it finds itself in the line of vision before the pupil, then the person acting as controller establishes the fact that the pupil width varies rhythmically with a frequency from 1 to 2 hertz. The patient himself notices the pulsation of the disk image. Naturally, oscillations in the pupil width occur without any changes being made in the lighting of the room. The explanation is relatively simple: "Normally only about one half of the change is compensated for by the reaction of the pupil. If, however, the centre of the pupil is shaded by the circular aperture, then the change in the size of the pupil has a much stronger effect. As will be noted, even after a jump in brightness the light can be fully, i.e. at 100 per cent, prevented from entering the pupil—except for diffused light. Consequently, the effect of the pupil's reaction, and hence the gain in the pupil's regulating circuit, is increased to an extraordinary degree by the central apertures. Whereupon this increase in the gain gives rise to the self-excited oscillations." (B. Hassenstein, 1967)

This clear example of a biological regulator goes to illustrate how effective the application of cybernetic approach can be in the treatment of biological phenomena.

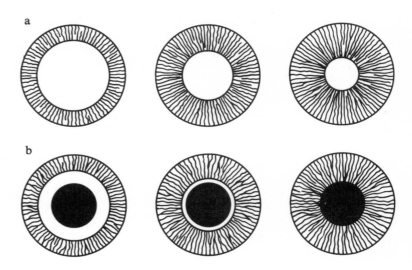

180   The pupil with central stop for producing self-excited pupil oscillations.

a—pupil without central stop. At an intensity of 1:4:16 (from left to right) the pupil area alters in the ratio 4:2:1 (50 per cent correction!).

b—if there is a central stop in front of the pupil, then the light stream into the eye is substantially more powerfully affected by pupil reaction, i.e. theoretically in the ratio 4:0.9:0. Consequently, the central stop enhances the effect of a pupil diminution or extension, and thus leads to oscillations in the pupil width. After B. Hassenstein.

The authors' aim has been to bring home to the reader the fruitfulness of a physical, technical and cybernetic approach to animate nature. This was done by giving twenty-three examples of phenomena, interesting and often inexplicable to the layman, from the vegetable and animal kingdoms. The authors' guiding principle in making their selection was, among others, that the basic data necessary for understanding should not exceed the given literary scope. In this connection let it be enough to name the human brain whose operations of data processing and storing as part of information bionics and neurobionics occupy a central position in bionic research.

Nowadays bionics experts throughout the world are trying to make the findings acquired in the sphere of biology technically and commercially applicable to Man. Thus, for instance, membranes with their all-embracing significance for metabolism in plants and animals, or else the cold light of various self-lucent animals, have become objects of intensive research. In many a field successes are emerging, or have already been achieved. For example, membranes are being employed in salt-water treatment; brain research promises to bring major clues for automatic dictating machines capable of converting the spoken word directly into characters, and for translating machines. In view of their extreme sensitiveness coupled with smallest dimensions and almost negligible power input, sensory organs serve as models for technical primary elements. Finally, let us mention a particularly simple and clear instance of an application of bionic findings: the slide zip-fastener.

Biology, the modern natural science, is on the advance. It is more than likely that towards the close of the present century biology will have acquired a similar degree of importance for human society as that held in the last fifty years by physics. New and new independent disciplines have been emerging out of biology. Let us just mention biophysics, biotechnology, molecular biology, biochemistry or bionics itself, to name only a few. All of these are penetrating more and more deeply into the structure and functioning of living organisms, daily unravelling new mysteries. This body of knowledge must and will be more intensively utilized by the engineer in the decades to come.

As for the extensive Bibliography, this was intended by the authors not so much as a list of references; it should much rather pave the way for the interested reader to further documentation.

Bibliography

Alexander, W.B.: *Die Vögel der Meere*. Hamburg and Berlin 1959

Antonov, P.: *Sportsprünge mit dem Fallschirm*. Berlin 1959

Beier, W. and G. Glass: *Bionik. Eine Wissenschaft der Zukunft*. Leipzig, Jena, Berlin 1968

Berndt, P. and W. Meise (Editors): *Naturgeschichte der Vögel*. Volume One, Stuttgart 1959

Bowsher, D.: *Einführung in die Anatomie und Physiologie des Nervensystems*. Bern, Stuttgart and Vienna 1973

Buch, H. and D. Strüber: *Abenteuer Fallschirmspringen*. Berlin 1973

Burkhardt, D., W. Schleidt and H. Altner (Editors): *Signale in der Tierwelt. Vom Vorsprung der Natur*. Munich 1966

Danert, S.: *Familie Palmen, Palmae oder Arecaceae*. In: Urania-Pflanzenreich. Höhere Pflanzen 2. Leipzig, Jena, Berlin 1973

Deckert, K.: *Familie Exocoetidae—Fliegende Fische*. In: Urania-Tierreich. Volume "Fische, Lurche, Kriechtiere". Leipzig, Jena, Berlin 1967

Drischel, H.: *Biologische Rhythmen*. Berlin 1972

Drischel, H.: *Einführung in die Biokybernetik*. In the series "Moderne Biowissenschaften". Vol. 8. Berlin 1972

Eck, B.: *Technische Strömungslehre*. 7th Edition. Leipzig, Heidelberg and New York 1960

Eisentraut, M.: *Aus dem Leben der Fledermäuse und Flughunde*. Jena 1957

Eisentraut, M.: *Der Winterschlaf mit seinen ökologischen und physiologischen Begleiterscheinungen*. Jena 1956

Fischer, A.B.: *Laboruntersuchungen und Freilandbeobachtungen zum Sehvermögen und Verhalten von Altweltgeiern*. In: Zoologische Jahrbücher. Abt. für Systematik, Ökologie und Geographie der Tiere. 96. Jena 1969

Franz, E.: *Der Flug der Insekten*. Bericht der Senckenbergischen Naturforschenden Gesellschaft. Vol. 89. No. 9/10. Frankfort-on-the-Main 1959

Frisch, K. von: *Tanzsprache und Orientierung der Bienen*. Berlin, Heidelberg and New York 1965

Gwinner, E.: *Orientierung*. In: Schüz, E. (Ed.): *Grundriss der Vogelzugskunde*. Berlin and Hamburg 1971

Hassenstein, B.: *Biologische Kybernetik. Eine elementare Einführung*. Jena 1967

Helmcke, J.G. and F. Otto: *Lebende und technische Konstruktionen*. In: Deutsche Bauzeitung. No. 11. 1962

Herforth, L. and H.M. Winter: *Ultraschall*. Leipzig 1958

Hermann, F.: *Meeresbiologie. Eine Einführung in die Probleme und Ergebnisse*. Berlin 1965

Hertel, H.: *Struktur · Form · Bewegung*. Mainz 1963

Herter, K.: *Der Temperatursinn der Tiere*. Lutherstadt Wittenberg 1962

Herzog, K.: *Anatomie und Flugbiologie der Vögel*. Jena 1968

Heynert, H.: *Einführung in die allgemeine Bionik*. Berlin 1972

Hoffmann, K.: *Temperaturcyclen als Zeitgeber der circadianen Periodik*. In: Zoologischer Anzeiger. 32. Supplement. Leipzig 1969

Hoffmann, K.: *Kompass und "innere Uhr" der Zugvögel*. In: Burkhardt, D., Schleidt, W., Altner, H. (Editors): *Signale in der Tierwelt*. Munich 1966

Informatik. In: Nova Acta Leopoldina. Vol. 37. 1. No. 206. Leipzig 1972

Jacobs, W.: *Fliegen · Schwimmen · Schweben*. Berlin 1938

Kaestner, A.: *Lehrbuch der Speziellen Zoologie*. Vol. I, Part 1, 2nd edition. Jena 1965

Kilias, R.: *Unterabteilung Coelenterata oder Radiata—Hohltiere*. In: Urania-Tierreich. Volume "Wirbellose Tiere 1". Leipzig, Jena, Berlin 1967

Kilias, R.: *Ordnung Siphonophora*. Ibidem

Klausewitz, W.: *Segelflieger der Hochsee*. In: Natur und Museum. Bericht der Senckenbergischen Naturforschenden Gesellschaft. Vol. 101. Frankfort-on-the-Main 1971

Klausewitz, W.: *Fliegende Tiere des Wassers*. In: Natur und Volk. Bericht der Senckenbergischen Naturforschenden Gesellschaft. Vol. 89. No. 9/10. Franfort-on-the-Main 1959

Kraismer, L.P.: *Bionik. Eine neue Wissenschaft*. Leipzig 1967

Kramer, W.O.: *Boundary Layer Stabilisation by Distributed Damping*. In: Journ. Am. Soc. Naval Engineers. 72. 1960

Kramer, W.O.: *The Dolphins' Secret*. In: New Scientist. 7. London 1960

Krompecher, St.: *Form und Funktion in der Biologie*. Leipzig 1966

Krüger, W.: *Bewegungstypen*. In: Handbuch der Zoologie. 8th Vol. 15th Printing. Berlin 1958

Krumbiegel, I.: *Biologie der Säugetiere*. Vol. 1. Krefeld 1954

Kruse, P.W., L.D. McGlauchlin and R.B. McQuistan: *Elements of Infrared Technology. Generation, Transmission and Detectation*. I. New York 1963

Kummer, B.: *Bauprinzipien des Säugetierskelettes*. Stuttgart 1959

Kummer, B.: *Biomechanik des Säugetierskeletts*. In: Handbuch der Zoologie. 8th Vol. 24th Printing. Berlin 1959

Kybernetik, Brücke zwischen den Wissenschaften. Frankfort-on-the-Main 1964

Lorenz, K.: *Der Vogelflug*. Pflullingen 1965

Lüthke, A.: *Die Uhr. Von der Sonnenuhr zur Atomuhr*. Düsseldorf 1958

Luca, G.G.: *Körper—Rhythmen. Die Uhr in uns geht ganz genau*. Hamburg 1973

Matauschek, J.: *Einführung in die Ultraschalltechnik*. Berlin 1962

Mauersberger, G.: *Familie Trochilidae—Kolibris*. In: Urania-Tierreich, Volume "Vögel". Leipzig, Jena, Berlin 1969

Mertens, R.: *Fallschirmspringer und Gleitflieger unter den Amphibien und Reptilien.* In: Natur und Volk. Bericht der Senckenbergischen Naturforschenden Gesellschaft. Vol. 89. No. 9/10. Frankfort-on-the-Main 1959

Miller, R.C.: *Das Meer.* Munich 1963

Nachtigall, W.: *Biotechnik. Statische Konstruktionen in der Natur.* Heidelberg 1971

Nachtigall, W.: *Gläserne Schwingen.* Munich 1968

Nachtigall, W.: *Die Kinematik der Schlagflügelbewegung von Dipteren etc.* In: Zeitschrift f. vergleich. Physiologie. 50. 1966

Nachtigall, W.: *Konstruktionsmorphologie: Funktionelle Gestaltung biologischer Stützelemente und Trägerstrukturen.* In: Verhandlungen der Deutschen Zoologischen Gesellschaft. Stuttgart 1973

Nachtigall, W.: *Patente der Natur.* In: Bild der Wissenschaft. 12. 1975

Nachtigall, W.: *Vogelflügel und Gleitflug. Einführung in die aerodynamische Betrachtungsweise des Flügels.* In: Journal für Ornithologie. 116. 1975

Natuschke, G.: *Heimische Fledermäuse.* Lutherstadt Wittenberg 1960

Norman, J. R.: *Die Fische.* Hamburg and Berlin 1966

Norman, J. R. and F. C. Fraser: *Riesenfische, Wale und Delphine.* Hamburg and Berlin 1963

Nürnberg, A.: *Infrarot-Photographie.* Halle 1957

Nusshag, W.: *Lehrbuch der Anatomie und Physiologie der Haustiere.* 7th edition. Leipzig 1966

Oehme, H.: *Untersuchungen über Flug- und Flügelbau von Kleinvögeln.* In: Journal für Ornithologie. 100. No. 4. Berlin 1959

Oehme, H.: *Vergleichende Untersuchungen an Greifvogelaugen.* In: Zeitschrift für Morphologie der Tiere. 53. 1964

Otto, F.: *Biologie und Bauen II.* In: Natur und Museum. Bericht der Senckenbergischen Naturforschenden Gesellschaft. Vol. 103. No. 4. Frankfort-on-the-Main 1973. Part I in No. 3

Patzelt, O.: *Naturkonstruktionen als Beispiele und Anregungen.* In: Deutsche Architektur. No. 12. 1969

Patzelt, O.: *Wachsen und Bauen—Konstruktionen in Natur und Technik.* Berlin 1972

Petzold, H.-G.: *Rätsel um Delphine.* Lutherstadt Wittenberg 1973

Petzsch, H.: *Familie Cynocephalidae—Pelzflatterer.* In: Urania-Tierreich. Volume "Säugetiere". Leipzig, Jena, Berlin 1966

Petzsch, H.: *Unterfamilie Pteromyinae—Flughörnchen.* Ibidem

Poniz, D.: *Die biotechnischen Tendenzen der Formgebung in der Baukonstruktion.* In: Bauwelt. 42. No. 58

Raths, P. and G.-A. Biewald: *Tiere im Experiment. Ergebnisse und Probleme der Tierphysiologie.* Leipzig, Jena, Berlin 1970

Rensing, L.: *Biologische Rhythmen und Regulation.* Jena 1973

Röhler, R.: *Biologische Kybernetik. Regelungsvorgänge in Organismen.* Stuttgart 1974

Sachsse, H.: *Einführung in die Kybernetik.* Reinbeck near Hamburg 1974

Salomonsen, G.: *Vogelzug.* Munich 1966

Salzmann, G.: *Segelfliegen.* Berlin 1958

Samal, E.: *Grundriss der praktischen Regelungstechnik.* Munich 1967

Schiele, M.: *Segelfliegen.* Stuttgart 1963

Schmidt, H. A. F.: *Flugzeuge aus aller Welt.* Vol. 1—4. Berlin 1974

Schmidt-König, K.: *Über die Navigation der Vögel.* In: Die Naturwissenschaften. 60. Göttingen 1973

Schmidt-König, K.: *Ein Versuch, theoretisch mögliche Navigationsverfahren von Vögeln zu klassifizieren und relevante sinnesphysiologische Probleme zu umreissen.* In: Verhandlungen der Deutschen Zoologischen Gesellschaft. Stuttgart 1970

Schnitzler, H.-U.: *Die Echoortung der Fledermäuse und ihre hörphysiologischen Grundlagen.* In: Orientierung der Tiere in Raum. Part 1: *Sinnes- und neurophysiologische Grundlagen.* Jena 1973

Schubert, E.: *Physiologie des Menschen.* Jena 1971

Slijper, E. J.: *Riesen und Zwerge im Tierreich.* Berlin and Hamburg 1967

Slijper, E. J.: *Riesen des Meeres. Eine Biologie der Wale und Delphine.* Berlin, Göttingen and Heidelberg 1962

Steinbacher, J.: *Der Flug der Vögel.* In: Natur und Volk. Bericht der Senckenbergischen Naturforschenden Gesellschaft. Vol. 89. No. 9/10. Frankfort-on-the-Main 1959

Struve, W.: *Die Eroberung der Luft in der Geschichte der Tierwelt.* Ibidem

Suworow, J. K.: *Allgemeine Fischkunde.* Berlin 1959

Tembrock, G.: *Biokommunikation. Informationsübertragung im biologischen Bereich.* Part 1. Berlin, Oxford and Braunschweig 1971

Thenius, E. and H. Hofer: *Stammesgeschichte der Säugetiere.* Berlin, Göttingen and Heidelberg 1960

Tischner, H.: *Ortungsbiologie.* In: Bild der Wissenschaft. February 1965

Tributsch, H.: *Fischen im Sturzflug.* Ibidem. November 1974

Weber, H.: *Grundriss der Insektenkunde.* 4th edition. Jena 1966

Wiener, N.: *Kybernetik. Regelung und Nachrichtenübertragung im Lebewesen und in der Maschine.* Düsseldorf and Vienna 1965

Wiener, N.: *Mensch und Menschenmaschine.* Frankfort-on-the-Main and Bonn 1964

Wissmann, G.: *Geschichte der Luftfahrt von Ikarus bis zur Gegenwart.* Berlin 1970

Wittke, G.: *Wärmehaushalt der Säugetiere und der Vögel.* In: Fortschritte der Zoologie. Vol. 18. 2nd Printing. Jena 1967

Index

ADN, Berlin 6
Birnbaum, Halle 78
Christiansen, Copenhagen 65, 151
Deutsche Fotothek Dresden 4, 45, 46, 112, 165
Deutscher Bilderdienst 96
Deutsches Museum, Munich 75, 76, 158
Florian, Leipzig 37
Friedrich, Leipzig 80
Grossauer/ZEFA, Düsseldorf 26
Halin/ZEFA, Düsseldorf 104
Hosking, London 100, 101, 150
Kabisch, Leipzig 35
Keystone Pressedienst Martin KG, Hamburg 36, 38, 48
Kolbe, Zerbst 62
Konrad, Leipzig 163
Kühlmann, Berlin 23
Lange, Leipzig 21, 22
Lindner/ZEFA, Düsseldorf 177 a
Marko/VEB Deutscher Landwirtschaftsverlag, Berlin 139
Moll, Waren/Müritz 98
Mönch/Okapia, Frankfort-on-the-Main 133 a
Müller, Leipzig 5, 79, 136
Nowosti Press Agency 47
Okapia, Frankfort-on-the-Main 58, 61, 97, 99, 152, 159, 161
Robiller, Weimar 113 a–c, 148
Scheithauer, Bad Aibling 103
H. A. F. Schmidt-Archiv/transpress Verlag für Verkehrswesen, Berlin 102
Schober, Leipzig 133 b–d
Schröder, Stralsund 105, 132, 149
Staatlicher Mathematisch-Physikalischer Salon, Dresden 140, 141
Thau/ZEFA, Düsseldorf 49
Tylinek, Prague 59, 60
VEB Deutscher Landwirtschaftsverlag, Berlin 138
Vogel, Leipzig 7, 17, 18, 19, 20, 24, 25, 63, 64, 71, 74, 77, 114, 134, 135, 160, 162, 164, 166, 177 b–d
Wittekind/VEB Deutscher Landwirtschaftsverlag, Berlin 137

From: Salomonson: *Vogelzug*. Munksgaard International Publishers Ltd. 146
From: Schüz: *Grundriss der Vogelzugskunde*. Paul Parey, Berlin 145
From: Slijper: *Riesen und Zwerge*. Paul Parey, Hamburg 15